Copernicus: A Very Short Introduction

VERY SHORT INTRODUCTIONS are for anyone wanting a stimulating and accessible way into a new subject. They are written by experts and have been translated into more than 40 different languages.

The series began in 1995 and now covers a wide variety of topics in every discipline. The VSI library now contains more than 450 volumes—a Very Short Introduction to everything from Indian philosophy to psychology and American history—and continues to grow in every subject area.

Very Short Introductions available now:

Available soon:

For more information visit our website

www.oup.com/vsi/

Owen Gingerich

COPERNICUS

A Very Short Introduction

OXFORD
UNIVERSITY PRESS

Oxford University Press is a department of the University of Oxford.
It furthers the University's objective of excellence in research, scholarship,
and education by publishing worldwide. Oxford is a registered trade mark of
Oxford University Press in the UK and certain other countries.

Published in the United States of America by Oxford University Press
198 Madison Avenue, New York, NY 10016, United States of America.

Library of Congress Cataloging-in-Publication Data
Names: Gingerich, Owen.
Title: Copernicus : a very short introduction / Owen Gingerich.
Description: Oxford ; New York, NY : Oxford University Press, [2016] |
Series: Very short introductions | Includes bibliographical
references and index.
Identifiers: LCCN 2016003101 | ISBN 9780199330966 (pbk. : alk. paper)
Subjects: LCSH: Copernicus, Nicolaus, 1473–1543—Popular works. |
Astronomers—Poland—Biography. | Astronomy—History—16th century.
Classification: LCC QB36.C8 G458 2016 | DDC 520.92—dc23
LC record available at http://lccn.loc.gov/2016003101

3 5 7 9 8 6 4 2

Printed in Great Britain
by Ashford Colour Press Ltd., Gosport, Hants.
on acid-free paper

Contents

List of illustrations

Acknowledgments

Special thanks go to Nancy Toff at Oxford University Press, who invited me to write this book, building on our collaboration on my Oxford Portraits in Science volume written a decade ago jointly with James MacLachlan, *Nicolaus Copernicus: Making the Earth a Planet*. James Voelkel, Richard Kremer, Jay Pasachoff, and Naomi Pasachoff have helped to improve the text by astute suggestions, for which I am forever grateful. Appreciation also goes to James Dobreff, who assisted by checking several of my Latin translations. Finally, a thankful acknowledgment to Noel Swerdlow for his two foundational works for modern Copernican studies, which are cited in the list of further readings at the end of this volume.

I take the opportunity here to record a note about the geographical place names in my text. In general I have used modern Polish names, and with the first use I have given in parentheses the sixteenth-century equivalent. The major exception is Frauenburg, Copernicus's professional base. Using the sixteenth-century name rather than today's Frombork is a reminder that it was and is a cathedral town with a major edifice for "Our Lady." Another, minor exception is Cracow, which joins Rome, Munich, and other international cities in having English equivalents.

Prologue

In or around 1510 Nicolaus Copernicus, one of the sixteen directors of the northernmost Catholic diocese in Poland, invented the solar system.

Wait a minute! you say. Wasn't the sun always in the middle of the planets?

But that wasn't the way everyone else thought about it. Farmers, professors, priests, and school children all assumed the earth was solidly fixed in the middle of the cosmos. Every day the sun and stars revolved around the earth. The sun also moved, more slowly, in a path against the more distant stars so that it was higher in the sky in the summer and much lower in winter.

Even when Copernicus's epoch-making book *De revolutionibus orbium coelestium* (*On the Revolutions of the Heavenly Spheres*) was finally published in 1543, very few readers imagined that the newfangled cosmology was a physically real description of the universe. Rather, it was a recipe book on how to calculate the positions of the planets.

Several generations passed, and only a handful of astronomers took the heliocentric cosmology seriously. The time-honored physics of Aristotle still ruled. In it the heaviest element, earth, automatically fell to the center of the universe, and fire rose to the top of the terrestrial elements. But Aristotelian physics was under challenge for anyone contemplating the implications of the Copernican system. In 1621 the poet John Donne expressed the confusion in rhyming lines:

> And new Philosophy calls all in doubt,
> The Element of fire is quite put out;
> The Sunne is lost, and th'earth, and no mans wit
> Can well direct him where to look for it.

Following the publication of his *De revolutionibus,* Copernicus's ideas would take a century and a half to gain a majority of educated followers. This *Very Short Introduction* is not the story of how the sun-centered cosmology won the center stage—the Copernican Revolution. That would take a volume at least twice this size, with an account of Kepler and Galileo, and finally of Isaac Newton. Here is the story of the busy churchman (not a priest), Nicolaus Copernicus, himself the competent mathematician and far-seeing cosmologist who helped revolutionize the modern world.

Chapter 1
Copernicus, the young scholar

Nicolaus Copernicus was born in Torun, Poland, on February 19, 1473, at 4:48 P.M. Apart from this stunningly accurate time (which needs to be taken with the proverbial grain of salt), essentially nothing is known about his childhood. He had an older brother and two older sisters. His father, a well-off merchant in this Hanseatic city, died when he was ten years old, and his maternal uncle, Lucas Watzenrode, took over responsibility for the boys.

Uncle Lucas had attended Poland's premier university, the Jagiellonian University in Cracow, and had subsequently undertaken graduate studies in Italy. He encouraged his young nephews to do the same. Meanwhile, he made great strides in ecclesiastical politics and in 1489 became bishop of Varmia, the northernmost Catholic diocese in Poland. Varmia was a tract of 2,000 square miles to the northeast of Torun, almost entirely surrounded by the part of Prussia still ruled by the German Teutonic Knights. The cathedral church of Varmia was located in Frauenburg (Frombork) on the Baltic coast, but the bishop's palace was situated in Lidzbark, forty miles southeast of Frauenburg. Being the bishop of Varmia was comparable to being governor of the region.

As the chief administrator, Uncle Lucas had considerable power over church appointments, and he could thereby enhance his

NICOLAUS COPERNICUS
TURENÆUS BORUSSUS MA-
THEMATICUS.

1. This woodblock print of Copernicus was created from his self-portrait. The lily of the valley that Copernicus holds is a symbol for Torun, his birthplace.

family fortunes; perhaps his bright young nephew could follow in his footsteps. In 1491 Nicolaus matriculated at the university in Cracow, a royal city along the Vistula River about 350 river miles south of Torun. Poland's strong Jagiellonian kings had fostered studies in the arts and sciences to compete with the older cultures in France and Italy. In particular, the Jagiellonian University in Cracow had scarcely any rival in all northern Europe in the study of astronomy, for it had not one but two professors of astronomy.

Education at the university was still closely tied to the seven liberal arts. At the lowest level were the three topics constituting the trivium (which has given rise to the word *trivial*): grammar, logic, and rhetoric. Young Nicolaus had presumably mastered these before arriving in Cracow, allowing him to concentrate on the four mathematically oriented topics of the quadrivium: arithmetic, geometry, music, and astronomy. Precisely which classes he took are unknown, but in his first semester the university's astronomy course used as a text the most famous introductory astronomy from the late Middle Ages, Sacrobosco's *De sphaera*, and undoubtedly he listened to these lectures.

The printing of books was then a comparative novelty, having begun in Germany only a few decades earlier. Sacrobosco's *De sphaera* in 1472 was the earliest printed astronomy text, which went on to hold the all-time astronomy record of more than two hundred editions. No doubt Copernicus had access to one or more of these editions, although soon they would prove too elementary to hold his attention. Rather, he would find a more advanced reference indispensable for his studies. This was a book of tables, composed in Paris early in the thirteenth century and named after King Alfonso of Spain. The *Alfonsine Tables* allowed an astronomer to calculate positions of the planets without going back to the basic geometry of the ancient geocentric system. The work was first printed in 1483, and Copernicus acquired the second edition, published in Venice in 1492.

In the winter of 1492 the university offered lectures on Euclidean geometry. By now the teenaged Copernicus had begun his lifelong love affair with mathematical astronomy. He obtained a copy of Euclid's *Geometry*, which had been printed for the first time in 1482. In those days books were sold as unbound stacks of paper, so Nicolaus would have sent the pages to a local binder. Because both his Euclid and the *Alfonsine Tables* have a characteristic Cracow design on their bindings, they provide evidence for his early burgeoning interest in astronomy. (The books were captured in the Thirty Years War and are today in Sweden in the Uppsala University Library, where they were preserved as one destructive battle after another raged across Poland through the centuries.)

Soon after Copernicus enrolled as a student in Cracow, information arrived concerning an astonishing discovery. An Italian navigator, sailing under Spanish colors, had crossed the Atlantic Ocean apparently to the Orient. Eventually a New World was revealed, and this would play a role years later when Copernicus began to write his magnum opus, mentioning not Christopher Columbus but Amerigo Vespucci.

At the beginning of 1495 Copernicus left the Jagiellonian University without taking a degree, and what he did next remains unknown. He could have gone off to another university such as Leipzig, although this is merely idle speculation. But in 1495 there was good news: Uncle Lucas the bishop had nominated Nicolaus for an appointment in the Varmian diocese, as he did later for Nicolaus' older brother, Andreas. However, there were complications and the appointment hung in the balance for two more years.

Nevertheless, by October 1496 Nicolaus was in Italy, enrolled in the ancient university in Bologna, studying law, both civil and canon (church). Canon law was particularly important in that period because the pope in Rome exerted great authority across Europe in matters of state as well as of church. If Nicolaus was to become an effective church official back in Varmia, he had to pay close attention to the lectures in Bologna.

Even more useful for his future fame was the fact that he lodged with Domenico Maria de Novara, the university's astronomy professor. More of an assistant than a student, Copernicus helped with observations, read the latest works on astronomy, and no doubt took part in many conversations with his mentor. Novara issued annual astrological prognostications, and the apprentice astronomer must have learned about the theoretical issues agitating astrological theory in that epoch. Nevertheless there survives not a word from Copernicus as to what he thought about horoscopes and planetary influences. Quite possibly, under Novara's tutelage, he charted for himself the positions of the planets at his birth. For the horoscopic chart to be oriented correctly, it was necessary to know the birth time to within four minutes, something unlikely to have been recorded with such precision. Novara would have known one or two standard tricks for deducing the missing number. The only whiff of Copernican astrology comes with a remarkably precise birth time on a horoscope chart that survives in the Bayerische Staatsbibliothek in Munich, definitely not in Copernicus's hand.

As long as his uncle held the purse strings, Nicolaus had to be careful with his expenditures. When at last he heard the news that opposition to his becoming a canon in the Varmian cathedral chapter had finally been overcome, he was so confident he would get the post that he promptly went to a local notary public for a document to authenticate his claim. The notary drew up a document that entitled two representatives in Varmia to act on his behalf in collecting his income. Now he could begin to make his own financial decisions, but he soon realized that he had been a bit too hasty. The official confirmation of his appointment came only a couple of weeks later. He returned the document to the notary, who updated it to October 10, 1497. Now it was all legal.

By the summer of 1500 the twenty-seven-year-old Copernicus had completed four years in the faculty of law at Bologna. It was time for a celebration, and without taking examinations for a degree, Nicolaus and Andreas headed to Rome. This was a jubilee year in Rome, celebrating fifteen hundred years of Christianity. Pope Alexander VI spared no expense on opulent entertainment for the celebration. Among others who came south for the anniversary was a Catholic monk from Germany, Martin Luther, who was shocked by the wealth that the profligate pope had squandered on his party and on his illegitimate children. The consequences of this impression would rock Europe, and even entwine themselves into Copernicus's later years.

Four decades later, Copernicus recalled for a young Lutheran disciple, Georg Joachim Rheticus, that while in Rome he had lectured on mathematics to a large crowd of students, important men, and experts; unfortunately, nothing more is known about the occasion or about his visit to Rome.

Following his Roman holiday, Nicolaus returned to Varmia, to be formally installed as one of the sixteen canons at Frauenburg. He was not a priest, even though he took charge of one of the altars of the cathedral, but he did take minor orders, including a pledge of celibacy. Being a member of the cathedral chapter was equivalent to being on the board of directors; the cathedral had immense land holdings in the diocese, and in later years Copernicus would be in charge of collecting rents, settling disputes, and preparing defenses against marauding Teutonic Knights. One of the three major Catholic military orders, the latter had in the thirteenth century settled in Prussia following the Crusades. Eventually there was strife with the Kingdom of Poland; Torun (Copernicus's birthplace) had become part of Poland in a peace settlement of 1466. Varmia was almost entirely surrounded by the territories of the Teutonic Knights, so persistent problems remained, and would require Copernicus's attention in later years.

Meanwhile, near the end of July 1501 Nicolaus and his brother appeared at a meeting of the Varmian cathedral chapter in Frauenburg. They asked to be allowed to return to Italy for further study. The chapter readily granted Nicolaus a two-year extension because he "promised to study medicine with the intention of advising our most reverend bishop in the future, as well as members of our chapter as a healing physician." The chapter also granted Andreas permission to be absent "for studies." It is not known if Andreas ever completed those studies, but in 1502 he represented Varmia in Rome in a dispute with the Teutonic Knights.

In October 1501, Nicolaus enrolled as a medical student at the University of Padua, the leading faculty for medicine in all of Europe. Doctors in those days were taught that the various signs of the zodiac influenced different parts of the body, so they needed to know some astrology to establish when and where would be the optimum time for a bloodletting. Thus once again Nicolaus encountered astrology. Whether or not he ever used astrology in his treatments, he did learn the techniques in Italy.

At the end of his second year in Padua, it was time for Nicolaus to return to Varmia. Although he had not completed the third year required for a doctor of medicine degree, it would be a bad idea to go back empty-handed. The cathedral chapter would prefer that he had some degree to show that his six years in Italy were not wasted. However, a degree from Bologna or Padua would be very expensive, not only to pay the examiners but also for a festive banquet expected by his fellow students. Frugal Nicolaus found a way out. He could go to nearby Ferrara, where he had no friends needing to be entertained, and take the examination in canon law.

On May 31, 1503, at age thirty, Nicolaus became Dr. Copernicus, graduating in canon law from the University of Ferrara. He returned to Varmia before the end of the year. Never again would he travel far from his homeland in northern Poland.

Chapter 2
The architecture of the heavens

In his years at the universities, Copernicus learned three
important details about celestial motions that would provide
a fundamental foundation for his future astronomical work.

First, the sun moves eastward in an annual path around the sky,
tilted so that it travels higher in the sky in summer. Furthermore,
it moves faster in winter than in summer, thus lingering longer
in the summer, and summer is in fact modestly longer than
winter. (This applies to the northern hemisphere, but in
Copernicus's day, who cared about any imagined people below
the equator?)

Second, the planets also move eastward, in approximately the
same path as the sun, but occasionally (about once a year) they
slow down, stop, and move westward for a while, a phenomenon
called retrograde motion. Like the sun, they also move on average
faster in one section of the sky, different for each planet.

And third, the stars themselves move very, very slowly eastward.
Thus, in antiquity when the sun was at its highest point in
summer, it was in the constellation of Cancer (which gives rise to
the expression "Tropic of Cancer"). By the time of Copernicus,
when the sun was at its highest point in summer, it was in the
constellation of Gemini.

There was, of course, much more to be learned. As early as Aristotle (fourth century BC), evidence for the sphericity of the earth was in hand. The earth's shadow on an eclipsed moon revealed that the earth was round. Furthermore, northbound travelers noted the way the northern constellations moved higher in the sky. To these observations Aristotle added an even more powerful theoretical argument. For him, there were four terrestrial elements: earth, water, air, and fire. Their natural motions were either up or down, always in straight lines (unless forced into unnatural motions). Since the earth was at the center of the universe, earth or water would always fall straight toward the center and would gradually accumulate as a giant ball.

To the uneducated, the notion of a round earth must have been highly counterintuitive. Surely the earth was flat. Thus, when universities began to be founded, and with astronomy among the seven liberal arts, the Aristotelian arguments for a spherical earth became a common part of the curriculum. Furthermore, a new argument against a flat earth was added: a sailor in the crow's nest atop the mast of a ship could see distant land before an observer could see it from the deck below.

In Aristotle's spherical terrestrial world, earth, water, air, and fire lay below the spheres of the planets and stars, which were composed of a fifth element, the celestial aether. Unlike the terrestrial elements with their natural up-or-down motion, the aether's natural motion was always circular. A contemporary mathematician, Eudoxus, envisioned a nested set of spheres to carry the planets (which included the sun and moon). He was a brilliant geometer—much of Euclid's geometry actually came from Eudoxus—but there were fundamental shortcomings with his conception. A given nested sphere was always the same distance from the earth. The problem was that the moon is not always the same size in the sky. Of course there is the famous "moon illusion" when in October the full moon seen near the horizon looks huge, but that is just a physiological or psychological phenomenon and not a real difference in the moon's angular size. But when the moon is at its nearest, it does appear about 10 percent larger than when it is at the most distant part of its motion. Similarly, the planet Mars appears considerably brighter when it is directly opposite the sun in the sky than when, for example, it is at right angles to the sun.

Eudoxus' so-called homocentric spheres was a theory to be admired, but because a given sphere always had the earth precisely at its center, there was no way it could account for the differences of brightness of Mars or the delicate changes of the moon's apparent size. Nor could a combination of earth-centered spheres track accurately the observed motion of the planet Mars.

Meanwhile, to the east, in the Mesopotamian lands of the Tigris and Euphrates, scribes were recording long runs of approximate planetary positions, along with meteorological data. But those scribes never attempted to get more than rough positions. If you were to measure the position of Jupiter tonight, how would you specify where it was? The Babylonians had no imagined coordinate grid in the sky. They simply estimated where the planets were with respect to conveniently placed "normal stars." The positions they recorded for the planets were pretty rough, but they gradually built up a library of clay "diary" records. Because the sequence of these cuneiform tablets eventually extended over several centuries, it became possible to derive surprisingly precise periods for the planetary movements. Their scribes eventually ascertained that after a number of years, the planetary positions repeated with fair accuracy. Thus, for example, they discovered that Mars, which rounds the sky in approximately 2.1 years, will repeat its pattern every forty-seven years. This led them to numerical schemes with which they could predict the planetary positions with fair success.

This information about Mesopotamian astronomy was, however, completely unknown to Copernicus. The Babylonian records had eventually been lost in crumbling ruins as their empires were overrun and turned to dust; they were not recovered until major archaeological work began in the Middle East in the nineteenth century. In 1849 the British archaeologist Austen Henry Layard, excavating the Assyrian capital of Nineveh, found the library of the seventh-century BC king Sennacherib, and a few years later his assistant, in digging on the other side of the mound, found the library of his grandson, Ashurbanipal. More than thirty thousand clay tablets or fragments were shipped to the British Museum. In the years that followed, thousands more were found in Mesopotamia and also sent to London. Between November 1876 and July 1882 the museum received fourteen thousand more pieces from Babylon, unfortunately with no records as to precisely where they had been found. Among these later findings were hundreds of tablets with astronomical data, which in general could be astronomically dated, typically in the fourth century BC. Besides discovering that the planetary positions repeated after a certain number of years, they noticed that each planet moved faster in particular zones in the sky, and they learned how to predict when a planet would begin or end its retrograde motion.

By the second century BC, the Babylonian data were being passed to the Greek astronomers. In particular, Hipparchus, an astronomer from Rhodes, acquired the precious records of eclipses and of planetary cycles. Thus began systematic attempts in the Greek-speaking world to map the heavens and to determine the planetary positions. This merger of the Babylonian arithmetic data with the Greek ideas of geometric model building meant that astronomy was at the threshold of an entirely new era. It reached its highest level around AD 150 in the work of Claudius Ptolemy, a scientist working in Alexandria (Egypt) in a Greco-Roman environment.

Ptolemy's *Mathematike syntaxis*, more than any other book, convinced people that the seemingly complex motions of the heavens could be represented by a simple underlying geometrical description, one that afforded the possibility of predicting celestial events. This was a major milestone in the development of science. Unlike his Babylonian predecessors, who loved arithmetic, he would build a geometrical model of the cosmos. His epoch-making book is generally called by its Arabic name, *Almagest*, meaning "the greatest." What he achieved—as far as we know, for the first time in history—is to show how to convert specific observational data into planetary parameters for his planetary models. With his models he could construct unprecedented tables whereby the planetary positions could be calculated for any given time, past or future.

It would become Copernicus's challenge to duplicate this achievement with an underlying model, one supplying a marvelous unification that went beyond what Ptolemy had imagined.

In some manuscripts of the *Almagest* is written this aphorism: "I know that I am mortal by nature, and ephemeral; but when I trace at my pleasure the windings to and fro of the heavenly bodies, I no longer touch earth with my feet; I stand in the presence of Zeus himself and take my fill of ambrosia, food of the gods."

Copernicus, in a different age, but a deeply religious culture, would express it this way: "So vast, without any question, is the divine handiwork of the Almighty Creator."

Chapter 3
Copernicus's vision

The Ptolemaic system, created by Claudius Ptolemy around 150 AD, for the first time in history introduced a comprehensive geometrical structure from which the positions of the planets (including the sun and moon) could be calculated for any time, past or future. It was a monumental achievement, which held sway for fourteen hundred years. Unfortunately, the Ptolemaic system has suffered from bad press, which has given the false impression that the system was unnecessarily complicated.

Ironically, it is easier to understand what Ptolemy did by looking at Copernicus's heliocentric arrangement. Let us start with the sun-centered orbits of the earth and Mars. We can begin with a kindergarten version, with two circles centered on the sun. The goal is to find the direction from Earth to Mars for any specified moment. We can calculate where the earth is because we know how swiftly the earth moves around the sun (approximately 1° per day, since there are 360° in the circle, and about 365¼ days per year), and we can establish an arbitrary starting point. Likewise we can do the same for Mars, which takes about 687 days to round the sun. This arrangement is shown in the left half of the diagram. What we need to find is the direction of the line from the earth to Mars, EM, which can be gotten with some simple trigonometry if we know the radii of the two circles and the

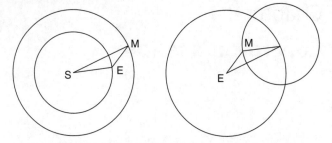

2. Left: the heliocentric orbits of Mars (M) and Earth (E) around the Sun (S) in the Copernican system. Right: the geocentric deferent circle for Mars with its epicycle.

angular positions of the radial lines. This, with details omitted, is how it works with the Copernican system.

Next, to see how it was done with Ptolemy, we switch the order of the circles, on the right half of the diagram. Notice that the direction of the line EM remains the same. But instead of having the circles centered on the sun, we have a geocentric system centered on the earth. Then, what started as the heliocentric orbit of the earth becomes an epicycle riding on a circle that began as the heliocentric orbit of Mars. In this system the large circle is called the *deferent*, literally the carrying circle because it carries the epicycle. We now have the basic structure of Ptolemy's epicyclic system (but keep in mind, this is the kindergarten version).

There is something quite wonderful about the heliocentric arrangement that is missing in the Ptolemaic arrangement, something that Copernicus noticed and that is one of the keys for understanding what he did. Look again at the heliocentric diagram, and imagine that the two planets are moving. The earth is moving faster, because it rounds the sun more quickly than slower-paced Mars does. Imagine how the line between them changes its orientation. Mars will appear to move backward in the sky for a few weeks as the faster-moving earth bypasses Mars. This is the famous retrograde motion of Mars, which happens every twenty-six months (just a little more than two years). Notice that this will always occur when Mars is directly opposite from the sun. The reason the retrogression takes place is perfectly obvious in the Copernican system. (If you don't find it perfectly obvious in the diagram, imagine you are in the fast lane on a superhighway, passing a slower car. With respect to the distant horizon, the slower car will appear to move backward as the overtaking proceeds.)

The same effect occurs in the Ptolemaic system as well, when Mars in its epicycle swings closest to the earth. But the connection to the sun is no longer obvious. Copernicus's student Rheticus, in introducing the heliocentric system, wrote that the motion of every planet is linked to the sun by a golden chain. Here is one of the golden chains. In the Copernican system, the sun is always on the opposite side of the earth from Mars when the retrogression takes place.

Now it is time to graduate from kindergarten. In this simple model, all of the retrograde loops should be identical. But in reality some are twice as long as those in the part directly opposite. Clearly some additional complication is required, in both the Copernican and the Ptolemaic arrangements. Furthermore, as Copernicus knew from his undergraduate studies in Cracow, each planet moves faster on average in one part of the sky than on the other side. For example, for Mars to move through the half of the sky centered on the zodiacal sign Capricorn requires on average 365 days, but in the other half of the sky only 320 days (approximately). Ptolemy also knew this, as did the ancient Babylonians before him. How can these phenomena be accommodated in a geometrical model?

If the deferent is set off-center from the earth, then the epicycle will appear to move more quickly in the part closer to the earth.

This also makes the retrograde loop shorter. But Ptolemy discovered that he could not find an off-center position for the deferent that took care of both phenomena at the same time—the varying speed of the planet and the length of the retrograde loop. He could use one eccentric position of the orbit to calculate the varying lengths of the retrograde loop, but he had to double the amount of eccentricity to take into account the varying speed of the planet as it moved around the zodiac. How could the eccentricity be doubled without disturbing the length of the retrograde loop?

Ptolemy's invention was very clever, though the Islamic astronomers who came a millennium later agonized that it was really cheating. What Ptolemy did was to choose another point, equal and opposite to the position of the earth.

This essentially doubled the eccentricity. The center of the epicycle revolved uniformly about this point, which was called the equant. Although Mars and the other planets moved uniformly about their equant points, since the deferents themselves were not centered on those points the speeds of epicycle centers actually varied. And that's what offended the Islamic astronomers. As later became apparent, Copernicus himself found the equant objectionable, though probably for another reason, which we will notice presently.

Ptolemy's solution, introducing the equant, was actually a brilliant stroke. It turned out to be a close approximation to what, centuries later, was Kepler's so-called law of areas and the ellipse. Ptolemy himself was a bit apologetic about the equant, because he knew the true-blue Aristotelians would find fault with a device that made the epicycles ride round on the deferent with a variable speed. Some centuries later the leading Islamic astronomers attacked the equant with an almost religious fervor. It was not greater precision of prediction they sought, but a purer mechanism for God's heavens.

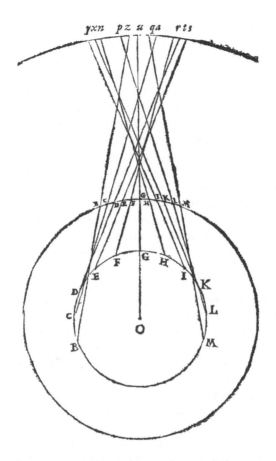

3. This diagram, drawn by Galileo, shows how retrograde motion is naturally explained in the heliocentric system. He shows the planets moving clockwise, the faster-moving earth on the inner circle, and the somewhat-slower-moving Mars on the middle circle. The lines of sight show how the position of Mars, as viewed from the earth, moves against the background of distant stars, the arc at the top of the diagram. At first, from positions B, C, D, and E, Mars appears to move clockwise against the starry background, but as the earth bypasses Mars (sight lines from F, G, and H), Mars appears to reverse its motion against the starry background.

19

Copernicus's vision, by contrast, was for a greater unity that answered a couple of unasked questions. The genius lay in asking those questions and in finding a profoundly satisfying answer that ultimately shattered age-old assumptions. In retrospect it all seemed so easy and obvious that it is hard to understand why it took the better part of two centuries for a majority of educated people to accept a major consequence of his grand unifying idea. But that major consequence seemed totally absurd. If the earth is spinning on its axis every twenty-four hours, the speed at the equator is easy to calculate: approximately 1,000 miles per hour. At that speed, wouldn't everyone and nearly everything else be spun right off into space? And, as Dr. Schöner, one of the leading astronomers in Germany, remarked in editing a manuscript from Johannes Regiomontanus, one of the greatest mathematicians of the fifteenth century, it would surely be much harder to walk west than to walk east.

So if this nonsense was the major consequence, what was the grand unifying idea that has established Copernicus as a memorable name in modern civilization? Ptolemy had made a great geometrical system whereby it was possible to compute the places of the planets for any time past or future. If you believed that the rhythms of the heavens controlled not only the rise and fall of great empires but also your love life, your children, your health, and your ultimate death, you would have to be grateful to Ptolemy for setting up such a precise calculating scheme. And you would have to be grateful to those Islamic calculators who continually worked over the tables to make the calculations simpler, even though they seldom if ever attempted to make the numbers more accurate.

Copernicus would have been happy to make a system that was immediately more accurate than Ptolemy's, but that was not the short-term result of his vision, although in a later generation under the pen of Johannes Kepler and with the treasury of observations from Tycho Brahe, there would be a major breakthrough in the accuracy of the predicted positions. But no, what makes Copernicus such a giant among astronomers was his invention of the solar system.

The emphasis is on both words, *solar* and *system*. For many years historians believed that Ptolemy never had a *system* uniting the separate mechanisms of all the planets, including the sun and moon, which also revolved about the Earth. Eventually, in the second half of the twentieth century, obscure manuscripts were found, some in Latin and some in Hebrew, showing that Ptolemy had built a sort of model wherein sets of deferents and epicycles were laid one after another so they wouldn't collide with each other. The epicycle of Mercury just cleared the moon's epicycle on one side and Venus's epicycle of the far side, and so on. But if the equants were to be simply represented in a mechanical model, there would have to be long arms, like the hands of a clock reaching from near the center of the model to all the planets. However, the pivot points for the physical linkages to Mars, Jupiter, and Saturn would fall on top of the mechanisms for the moon and Mercury, and this would have made a terrible mess of things if an actual mechanical model were required.

On the other hand, Copernicus's sun-centered system arranged the planets in an entirely unexpected and beautiful way. His system laid out the planets around the sun according to their periods, and something wondrous happened. Swift-moving Mercury had the smallest orbit, and lethargic Saturn had the largest, and the others all fit conveniently in the intervening space. This arrangement automatically and unambiguously arranged the planets according to their periods of revolution around the sun. Nothing bumped into anything else. (There was a potential problem with the equants, but we will return to that later, in the first appendix.) The powerful beauty of this unified arrangement must have stopped Copernicus in his tracks. It was too awesome to ignore. Even the book title that ultimately emerged from this discovery carried the message: "*On the Revolutions.*" It was the dance of the planets, now including the earth, all arranged according to their periods of revolution, that produced an ethereal aesthetic.

Furthermore, it now gave a logical explanation for the phenomenon of retrograde motion. Instead of being a capricious glitch of a planet every so often, it was tied, as by a golden chain, to the position of the sun, always (for the superior planets, Saturn, Jupiter, and Mars) when the planet was opposite the sun in the sky. And on the other hand, it explained why Venus and Mercury can never appear opposite the sun in the sky.

Unfortunately, Copernicus left behind almost no archival record of how or why he got the idea of arranging the planets in this radical formation. Had he heard about such an arrangement from an antique source? In his day, being too curious or inventive was not always seen as a virtue. Copernicus would have felt less vulnerable if he could show that others had preceded him. In fact, when he first drafted the front matter of his book, he included a brief reference to Aristarchus, a third-century BC Greek, which said in part, "Philolaus believed in the mobility of the earth and some even say that Aristarchus of Samos was of that opinion. But since such things could not be comprehended except by a keen intellect and continuing diligence, Plato does not conceal the fact that there were very few philosophers in that time who mastered the study of celestial motions." But Copernicus wanted to be more erudite, so he switched out the Latin quotation for a different one in Greek, and as it happened, Aristarchus's name wasn't in the Greek replacement.

Neither the original passage nor its replacement says much of anything about Aristarchan cosmology. Had Copernicus known more, he would surely have been happy to mention it, since he needed to muster all the support he could get for his own unorthodox views, and since he quotes with enthusiasm other possible geokineticists from antiquity with less reliable credentials.

But none of the vague references from ancient astronomers said anything about the fact that, when the planetary orbits are arranged around the sun, they automatically fall according to their period of revolution. As Copernicus would eventually say in his *Revolutions*, "In no other way do we find a sure harmonious connection between the size of the orbit and the planet's period of revolution." There is no question but that Aristarchus had the priority of the heliocentric idea. Yet there is no evidence that Copernicus owed him anything. As far as we can tell, both the idea and its justification were found independently by Copernicus.

Chapter 4

Canon days and the *Little Commentary*

It was payback time when the thirty-year-old Copernicus returned to Varmia in 1503 after his six years of graduate study in Italy. Uncle Lucas had pulled the strings to support the young man in Bologna and Padua. Now, with his newly minted doctorate in law and qualifications in medicine, Nicolaus was essentially an indentured servant working at the beck and call of the powerful bishop of the northernmost diocese in Poland, Lucas Watzenrode. This meant being assigned to the Bishop's Palace in Lidzbark Warminski, approximately forty miles to the east-southeast of Frauenburg.

What kept Dr. Copernicus busy in the six years he acted as his uncle's private secretary is only guesswork. We know he occasionally traveled on governance affairs in his uncle's retinue to Malbork (Marienburg) and Elbląg (Elbing), but otherwise the records, if there were any, are lost. We know that his uncle was seriously ill in 1507; Copernicus's knowledge as a physician was so useful that the Varmian chapter voted him a special bonus.

Beyond these scarcely recorded duties, one idiosyncratic project stands out. In 1509 Copernicus published a Latin translation of the Greek "moral, rustic, and amatory epistles" of Theophylactus Simocatta, a seventh-century Byzantine historian. The work was prefaced by a long Latin poem by his friend Lawrence Corvinus,

4. Varmia, the northernmost diocese in Poland, where Copernicus served as canon, was surrounded by the lands of the Teutonic Knights. This map shows the boundaries as they appeared around 1526.

who presumably arranged for its publication in Cracow. Why Copernicus chose this minor work to whet his nascent knowledge of Greek is puzzling, but the eighty-five epistles are amusing and short (many only five or six lines long) and were available in a recently printed Greek edition. In any event, his improved knowledge of Greek would serve him well when he needed to convert the dates of early Greek planetary observations into Latin equivalents.

Gradually Copernicus's passion for astronomy claimed a larger piece of his life, and late in 1510 he left the palace in Lidzbark for a residence in Frauenburg. If we could only sort through the waste paper from his many calculations made in the following decades, we could no doubt piece together parts of the process leading up to his radical heliocentric world view, a planet Earth spinning on its axis and speeding around the sun in an annual orbit. And how helpful it would be if every scrap contained a working date! But almost nothing of the sort survives. Nevertheless, two pages of random notes in one of his astronomy books supply tantalizing clues. The book was bound in Cracow, where he was an undergraduate, and is now preserved in Uppsala in Sweden. The first fascinating note is a short, cryptic, undated statement in highly abbreviated Latin. It is scrawled just after a few lines concerning two lunar observations made in Bologna in 1500:

> Mars superat numeratione plusque g ij
> Saturnus superatur a numero g 1½

These translate into:

> Mars surpasses the numbers by more than 2 degrees;
> Saturn is surpassed by the numbers by 1½ degrees.

Until large programmable computers became available in the 1960s, it would have been a formidable challenge to search for a time when the numbers in the standard tables of the day deviated from the actual positions by those amounts. And where was Jupiter in that lineup?

Copernicus did not have instruments of the sort that could give accurate positions against a hypothetical sky grid. But when planets were moving close to one another, a phenomenon called a conjunction, it was relatively easy to watch for the closest approach and mark the calendar for that date and time. Slow-moving Saturn and Jupiter come into conjunction approximately every twenty years. It had happened in 1484, when Copernicus was not yet a teenager. But surely in 1504 at the conjunction's repeat, Copernicus would have wanted to observe and record this relatively rare event. Unfortunately, the actual conjunction took place in May 1504, when those planets were too close to the sun to be seen. However, in the run-up to the conjunction, the swifter planet Mars passed close to Jupiter and then to Saturn; subsequently it went into retrograde motion and backed past Saturn and then Jupiter. When Mars resumed its direct motion at the beginning of February 1504, it quickly passed Jupiter and then Saturn. These observations posed a real puzzle, and it must have taken Copernicus quite some calculating before he found that the predictions for Jupiter were right on target, but for Saturn the predicted numbers were about 1½° too high and for Mars 2° too low.

We might have expected Copernicus to mention in his later work the discordance between these rare conjunctions and their predictions, but he passed over them in silence. In any event, they provide evidence that his interest in astronomy continued in the months following the return from Italy.

The second fascinating fragment in the Uppsala volume is even more intriguing, but so far it resists being dated. Quite possibly it comes from around 1510, after Copernicus left his position with the bishop. It comprises two groups of numbers, both of which refer to the size of the planetary spheres. In Ptolemy's *Almagest* the size of Mars' or Jupiter's or Saturn's deferent is the same, sixty parts, as they are treated individually and not as part of a system. The epicycle represents the comparative size of the sun's orbit if we are thinking geocentrically, but if we wish to make the monumental leap to a heliocentric system, then the epicycle represents the comparative size of the earth's orbit. In that case Copernicus would want the "epicycle" or the "eccentric" (as he calls it) to be a fixed standard size. So here is the two-stage process. Instead of sixty parts, Copernicus used ten thousand parts in the first table, getting these scaled-up numbers for the epicycles of the so-called superior planets:

	Deferent	Epicycle
Mars	10,000	6,583
Jupiter	10,000	1,917
Saturn	10,000	1,083

Now Copernicus wished to rearrange the role of the circles, making the second column a standard number. With the size of the earth's sphere arbitrarily chosen as twenty-five, the planetary deferents are suitably scaled as follows:

	Scaled deferent	Scaled eccentric
Venus	18	25
Mars	38	25
Jupiter	130 25	25
Saturn	230 5/6	25

The second table carries the step one stage further. Copernicus was exploring whether the epicycle could play the role of a standard sphere for the earth, and for that he needed to make all the eccentric numbers the same.

This was surely *not* the eureka moment when Copernicus conceived of the heliocentric universe. That remains cloaked in mystery. Nevertheless these are fragments from the process of reconstructing the planets into a system. By using 25 units as the standard size of the earth's sphere, he had calculated the comparative size of each planet's sphere. In other words, as we would say today, Copernicus had found the size of the orbit for each planet compared to the earth's orbit. There they stand in the table, 18 units for Venus, 25 implied for the earth, 38 for Mars, 130+ for Jupiter, and 230+ for Saturn.

In a way, it looks as if the heliocentric system is well at hand, for additional data in the tables not shown here tabulate the sizes of a pair of small epicyclets that Copernicus will use to replace Ptolemy's equant (see Appendix 1 for more details).

But the heavy lifting is still to come. It's not just a problem with the so-called equant. It's the whole philosophy of what holds up the sky. It's not so difficult for us to trace the steps of substituting the Ptolemaic epicycle to become the earth's orbit in a unified planetary system, but in so doing we are using familiar post-Newtonian concepts. The idea of orbits was barely in the vocabulary of sixteenth-century astronomers. But if Copernicus's predecessors and his contemporaries were not using orbits, how did they envision what kept the planets moving?

For thinking about the paths of planets, the most popular guide for early-sixteenth-century astronomers had been written around 1470 by Georg Puerbach, a professor at the University of Vienna. It was called *Theoricae novae planetarum*. It was not, as you might guess, a "new theory of the planets." It was a new "theorica" of the planets, where here the word means an arrangement or modeling of the planets' motions. The new theorica showed how zones of celestial aether could slide inside outer layers of aether to carry planets in eccentric paths. This was a fully Aristotelian concept, for in his physics the heavens were composed of a quintessential weightless aether whose natural motion was perfect circles, endless in their motions. There was no problem of the sky falling in for there was no natural downward proclivity. And because nature abhorred a vacuum, any celestial voids were simply filled with the fifth element, the quintessence.

For students in Western Europe, the Puerbachian spheres were a compelling new idea, though they were long known among the astronomers working in the Islamic world.

We have seen how the Ptolemaic circles for Mars could be transformed into a circle for Mars and one for the earth. Let us suppose that Copernicus had a sheet for each planet with its epicycle transformed into a circle for the earth. For Copernicus to stack all the planets into a uniform system, he had to make sure that all the circles for the earth were scaled to be the same size, as shown in the previous table. Copernicus had already done this numerically on the Uppsala manuscript page, with the results shown here for the size of the circular planetary paths, to which I have also added the approximate times required to go around the sun:

Mercury	8	88 days
Venus	18	225 days
Earth	25	365 days or 1 year
Mars	38	2 years
Jupiter	130 25/60	12 years
Saturn	230 50/60	30 years

Copernicus's brilliant sleight of hand enabled him to arrange the planets with Mercury nearest the sun. Mercury, the fastest one, sped around the sun four times in a year. Venus, with the next largest orbit, circled the sun in 225 days. And so on, out to Saturn with its almost thirty-year period. It was a powerful unification, too beautiful to lose. But it tore at the entire fabric of a neatly stacked plenum—literally, a full universe. Copernicus's arrangement left monstrous voids between the planets. Without a plenum universe with one crystal sphere sliding upon another, what would hold up the sky? How could the heavy earth be stable in a surround of weightless aether? One man's elegance was another's terror. It's no surprise that Copernicus had to think long and hard about what was to come, and how to buffer his system against criticism both academic and bucolic.

It was now time for Copernicus to write out a brief sketch about the new arrangement, something approximately the size of Puerbach's *New Theorica of the Planets*. His original paper explaining his new arrangement hasn't survived, but three early copies are known, one in Vienna, another in Stockholm, and the latest one to be recognized, in Aberdeen, Scotland. Apparently Copernicus didn't give his manuscript a title, but later owners of the copies labeled it *Commentariolus*, that is, *The Little Commentary*. Here is how it began:

> I understand that our predecessors assumed a large number of celestial spheres mainly in order to explain the apparent motion of the planets through uniform motion, for they thought it altogether absurd that a heavenly body should not always move uniformly in a perfectly circular figure. They had discovered by the arrangement and combination of uniform motions in different ways, it could be brought about that any body would appear to move to any position.
>
> Calippus and Eudoxus, who tried to achieve this by means of concentric circles, could not thereby account for everything in the planetary motion, that is, not only those motions that appear in connection with the revolutions of the planets, but also that the planets appear to us at times to ascend and at other times to descend in distance, which concentric circles in no way permit. And for this reason a preferable theory, in which the majority of experts finally concurred, seemed to be that it is done by means of eccentrics [= deferents] and epicycles.
>
> Nevertheless, the theories concerning these matters that have been advanced far and wide by Ptolemy and most others, although consistent with the numbers, also seemed quite doubtful, for these theories were inadequate unless they also conceived of certain

equant circles, on account of which it appeared that the planet never moves with uniform velocity either in its deferent sphere or with respect to its proper center. Hence a theory of this kind seemed neither perfect enough nor sufficiently in accordance with reason.

Thus when I noticed these defects, I often wondered whether perhaps a more reasonable model composed of circles could be found from which every apparent irregularity would follow while everything itself moved uniformly, just as the principle of perfect motion requires. After I had attacked this exceedingly difficult and almost insoluble problem, it at last occurred to me how it could be done...if some postulates are granted:

...

Third Postulate: All spheres surround the sun as though it were in the middle of all of them, so that the center of the universe is near the sun.

...

Sixth Postulate: Whatever motions appear to us to belong to the sun are not due to the motion of the sun but to the motion of the earth and our sphere with which we revolve around the sun just as any other planet. And thus the earth has then more than one motion.

Copernicus's opening volley was subtle and reserved. His third and sixth postulates were the fundamental ones from which other postulates followed, including the seventh, which, however, was one of the strongest arguments on behalf of the heliocentric system.

Seventh Postulate: The retrograde and direct motion that appear in the planets belong not to them but to the motion of the earth. Thus the motion of the earth alone suffices to explain a considerable number of apparently irregular motions in the heavens.

Immediately after presenting his radical postulates, he turned to his remarkable finding: that in this arrangement the fastest-moving planet, Mercury, circled closest to the sun, the next swiftest, Venus, took the next position, then the earth, and so on. Thus the earth became a planet, and the group became the solar system. But he did not flaunt this.

The remainder of the *Commentariolus*, in fact the bulk of it, outlines in swift paragraphs the structure of the zones for the sun, moon, and other planets, explaining the arrangements used both for predicting their longitudes as they wheeled around the sky, and for their movements in latitude, that is, their short motions north or south of the sun's apparent path (the ecliptic). At the end of the *Commentariolus*, Copernicus boasted, "And so altogether, Mercury moves on seven circles, Venus on five, the earth on three and the moon moves about it on four, and finally Mars, Jupiter, and Saturn on five each. Therefore, taken as a whole, 34 circles are sufficient to represent the entire structure of the heavens and the entire ballet of the planets."

Many twentieth-century scholars assumed that Copernicus was exuberant because he had radically simplified the overburdened Ptolemaic system. Already in 1831 the astronomer John Herschel had described pre-Copernican astronomy as "'cycle on epicycle, orb on orb,'—till at length, as observation grew more exact, and fresh epicycles were continually added, the absurdity of so cumbrous a mechanism became too palpable to be borne." Or, as the 1969 *Encyclopaedia Britannica* put it, by the time of Alfonso in the thirteenth century, forty to sixty epicycles were required for each planet (!). These remarks are historical nonsense. In fact, the Ptolemaic system and the Alfonsine Tables required *fewer* circles than the *Commentariolus*. Copernicus was exuberant because in spite of the apparent complexity of celestial motions, the ballet of the planets could be represented by as few as thirty-four circles.

Precisely when did Copernicus formulate his *Commentariolus,* his first impression of the heliocentric cosmology? This is what all of his biographers have dearly wanted to know. Without getting a precise answer, they have found a *terminus ante quem*, that is, it must have been written by 1514 or earlier. In the Jagiellonian Library in Cracow there is a catalog of manuscripts and books owned by a sixteenth-century professor of medicine, Matthew of Miechow. It contains a brief entry dated May 1, 1514, for an item "asserting that the earth is moved while indeed the sun is at rest." As the eminent Copernican scholar Noel Swerdlow has written, "It is hard to believe that this could refer to anything other than the *Commentariolus* for then we would be faced with the remarkable possibility that there was someone other than Copernicus who wrote a description of the heliocentric theory that found its way to Cracow by 1514."

Arranging the earth and other planets around the sun now seemed to call for a new illustration showing the heliocentric arrangement. None of the early copies of the *Commentariolus* included a diagram, but eventually Copernicus supplied one. It is beautifully done because he was a skilled draftsman, but to modern observers who expect to see orbits, it is subtly misleading.

There was a difficulty at the very beginning. If Copernicus started to draw a scale model of the solar system, there would quickly be a problem. Suppose he chose 0.6 inch as the radius for Mercury's orbit around the sun. That same scale would place Earth at a comfortable 2½ inches from the sun, but Saturn's radius would fall at 23 inches, or an orbital diameter of nearly 4 feet. It would be hard to find a sheet of paper large enough. So it is not a surprise that no scale drawing survives from his study.

Copernicus's classical diagram is not a scale model of the heliocentric layout; nor does it show orbits. One might even say that it is a public relations view of the universe, with many distracting aspects suppressed. A casual glance at the diagram (as many readers have given it) leads to the conclusion that the planets are revolving about the sun in perfect circles, of course much simpler than the convoluted old Ptolemaic system. However, Copernicus was not as simple-minded as that! What he shows are not orbits, but cross-sections of the planetary spheres. These are the zones in which the planetary mechanisms lie, the eccentric circles. Notice in the picture that the earth's zone is big enough to encompass the moon's spheres as well. Notice also that

5. This diagram of the heliocentric system appears near the beginning of *De revolutionibus*. It shows not the orbits of the planets but the zones in which the eccentric orbits and the moon reside.

the planetary zones are packed tightly together, harking back to an earlier concept of a plenum universe. It is a good start for getting all the planets on the same page, albeit somewhat misleading.

But now, how to bring Ptolemy's equants into the picture? Remember that these are auxiliary circles the same sizes as the deferents but each with its own center point. In overlaying one planet's machinery over another to produce a unified picture with all six planets, the *centers* of the larger equant circles will fall in a jumble onto the spheres for Mercury and Venus. This could well be the reason that Copernicus found the equant so undesirable. How much more elegant matters would be if the each planet's equant could somehow be embedded within its spherical zone!

Happily, Copernicus found a clever way to do precisely this. Details as to how this worked are described in Appendix 1. With the required effects of the equants preserved, but the equant circles

35

themselves replaced by mechanisms bundled within the zones of each planet, the solar system was orderly and beautiful.

Although the details were too arcane for all but the specialists, Copernicus's diagram was an uncluttered advertisement for a totally new way to think about the universe. Nevertheless, he was reluctant to make it public. As he would later write, he didn't want to be hissed off the stage. That is precisely how he expressed it in the dedication he eventually wrote to Pope Paul III. Some years ago I became curious as to what Latin word Copernicus employed to express this. He used *explodendum*. The *Oxford English Dictionary* confirms that this is also the original, but now obsolete, meaning of the English word *explode*, "to be hooted off the stage." *Explode* didn't pick up the modern definition of "to blow up with a loud noise" until around 1700. Shakespeare never used the word despite its theatrical connotation. And in the end Copernicus was not exploded.

Chapter 5
Competing with Ptolemy

In 1515 the printing office of Peter Liechtenstein in Venice brought out a book much desired by Copernicus: a full Latin translation of Ptolemy's account of his geocentric cosmology. The original Greek title, *Mathematike syntaxis*, meant *Mathematical compilation*, but the Islamic astronomers called it *Almagest*—literally "the greatest"—and that was on the title page of the new publication. Copernicus knew the general outline of Ptolemy's book because a Latin epitome of the work had been published in Venice in 1496. That was the work of Georg Puerbach and Johannes Regiomontanus, two of the outstanding astronomers of the fifteenth century. They had learned Greek for the purpose of translating the *Almagest*, though in the end what they produced was not a full edition but an insightful and detailed commentary of Ptolemy's text.

With the new volume in hand, Copernicus must have quickly realized that his radical cosmology would need a comparable book to be taken seriously. He must also have understood that such a book would require a full set of observed planetary positions to gain comparable standing to Ptolemy's monumental work. A "full set" of observations means approximately four observations per planet—perhaps a shockingly small number, but having additional observations with their inevitable small errors would simply have confused things. However, not just any four observations for each planet were called for, but observations made at critical times with respect to the geometry of the planet's motion, which in some cases meant waiting for years to get just the right configuration. And thus it was that Copernicus spent more than two decades collecting the specific observations.

In his *De revolutionibus*, his magnum opus finally printed at the end of his life, Copernicus mentions three instruments he may have used. One was a simple type of sundial for measuring the altitude of the sun at noon. He said it could be made from a square of wood, though some stone or metal would be better, and he explained how it could be set up accurately in a north-south line. On one side a quarter circle was marked in degrees. Copernicus was not very explicit about specific observations he made with such a dial, though he reported that he had observed the times of the equinoxes for more than ten years. To determine the day of the equinox, when the sun is midway between its northernmost and southernmost travels (and therefore on the celestial equator), he had to mark the highest and lowest positions of the noon shadows during the year and bisect the angle between them. When the noon shadow fell on the bisection point, that was the time of the equinox.

Two additional complex instruments were used for the positions of the stars. Remember that these were naked-eye devices, for the telescope would not come until the following century. We know quite a bit about one of these instruments, called the Ptolemaic rulers or triquetum, because many years later the Danish astronomer Tycho Brahe sent an observer to Frauenburg to see what survived from Copernicus's time. His observer brought this instrument back to Denmark. Tycho was very proud of having the Ptolemaic rulers. He immediately wrote a heroic poem to celebrate them, and he described them in his own book on instruments.

The most complicated instrument discussed by Copernicus was called by Ptolemy an astrolabe. But today we refer to it as an armillary sphere, because what we now call an astrolabe is an entirely different sort of device, a flat set of brass plates representing the stars and the horizon for chosen latitudes. Copernicus may well have owned such an astrolabe, though he didn't mention it. He describes the armillary sphere as consisting of a series of nested rings, one set of which contained sights for setting on the stars or planets. It is a little frustrating that he doesn't tell us the size of his armillary sphere. He just says it can't be too large or it will be hard to handle, nor too small because the rings have to be divided into degrees and minutes. Perhaps his was 2 or 3 feet across—if indeed he owned one, for he never claims to have made any of his observations this way.

Approximately forty observations in two decades does not seem particularly burdensome, though he probably made many more that do not survive. Of the forty, he would use twenty-seven of them in his book. One of the "observations" of an opposition of Saturn was specified as noon on July 13, 1520. Obviously Saturn was invisible in the middle of the day, and the reported position must have been interpolated from nighttime observations before and after that specific date.

But how could Copernicus measure the position of a slowly moving planet? There is no coordinate grid marked on the sky, although the planets could be measured against nearby stars—providing the positions of the stars were known. Aye, there's the rub. There is continuous movement of the coordinate system with respect to the stars, known as precession of the equinoxes. In Copernicus's view the earth was spinning like a top, its axis lazily tracing out a large circle in the sky. Today the earth's pole points closely to the second-magnitude star Polaris. In his day, Polaris was about 3° from the celestial pole and was becoming a fairly good navigational star as the Age of Exploration began. Back in 150 AD Ptolemy's star table showed it to be 11° from the celestial pole, not even in the running to be a pole star.

Besides knowing the rate of precession, it was useful to have a starting point for the precession system, just as geographers eventually adopted the Greenwich Observatory in the environs of London as the starting point for the measurement of terrestrial longitudes. Copernicus's preference was to choose a particular star as the zero point for celestial longitudes, but he stood alone on that opinion. Astronomers then and now chose the invisible spot in the sky where the sun crosses from the southern half into the northern half of the sky, and this spot is slowly moving among the stars. Ptolemy thought its movement amounted to a degree per century, which is about 30 percent too slow. Ten centuries later it had become obvious a correction was required, but the Islamic astronomers didn't want to mess with the value of precession attributed to Ptolemy in his day, so they introduced a variable rate, which by the fifteenth century was getting obviously wrong.

Regiomontanus, who was deeply involved in translating Ptolemy's *Almagest,* was also the leading printer of scientific books in the first decades of printing with movable type. In addition he happened to be the best mathematician of the century. Furthermore, he was an observer as well. In 1465 he noticed that there was something appreciably wrong about the predicted position of Mars: "On the 26th day of April around the beginning of the night Mars was seen in a straight line with 1 and 2 of Sagittarius, or a little north of that line, and was 1½ times farther than the distance to the first star, whereas by computation Mars was at 6° of Capricorn." Without knowing the position of the first or second star in Sagittarius, the readers of his book would have been baffled as to how much discrepancy was involved. The problem was, how to find the numerical position of that star so the comparison could be made. A reliable fresh version of the precession was not readily available, and Regiomontanus, who was clearly clever enough to find it, simply had too many other projects on his hands.

This, then, was the starting problem for Copernicus. He had to get precession under control before he could get into serious details of the planetary motions. And we know he took the problem seriously, because in 1524 he wrote a long, technical letter that was essentially a demolishing review of a small book by a Nuremberg geometer and pastor, Johann Werner: *De motu octavae sphaerae* (1522). "The motion of the eighth sphere" was essentially the motion of precession. In Copernicus's view the eighth sphere was solidly fixed, while the earth was spinning round, but his radical heliocentric cosmology was well concealed in his rebuttal to Werner. Instead, in his "Letter against Werner," the Varmian stronomer demolished the Nuremberg scholar by showing an eleven-year error in his basic chronology. Unknown to Copernicus, however, Werner had died of the plague before his rebuttal circulated. Meanwhile, Copernicus also had other, cathedral business on his plate.

Once he had left his uncle's service in Lidzbark, in 1510 he had returned to Frauenburg (which was the working capital of Varmia). Pledged to celibacy and with the minor orders of a religious institution, he was on a board of directors comprising well-educated, upper-class members of Polish society. His fellow canons promptly elected him to the high post of chancellor, where he was responsible for drafting the documents and letters required in the conduct of the chapter's business. In 1516 he became administrator of rents and adjustor of land claims, an important task since the cathedral and palace were the largest landowners in Varmia, and all the canons would profit from Copernicus's wise handling of the tenant farmers. This position required his going to Olsztyn (Allenstein), about 50 miles southeast of Frauenburg, so he often lived there in the chapter's administrative castle. He performed these duties for three years and then again in 1521.

The rules of the cathedral chapter meant that as a canon he was entitled to a house in the town, beyond the walled perimeter of the Frauenburg cathedral grounds, and he was expected to keep three horses and two servants. Over the years he owned various residences. In 1514 he upgraded to a different house, and at the same time the records show that "Doctor Nicolaus" purchased eight hundred bricks and a barrel of chlorinated lime, presumably to add a flat observing platform near his new dwelling.

In addition Copernicus also obtained rooms in a tower in the wall of the cathedral quarters. This was particularly useful, because in January 1520 rampaging Teutonic Knights burned down the town of Frauenburg, destroying all the homes of the canons, including that of Copernicus. Presumably his work-in-progress was preserved in his cathedral complex quarters.

The burning of Frauenburg was the opening salvo that started an active war, and Copernicus necessarily took an active part in organizing the defenses in Varmia. A year later, however, Poland and the Teutonic Order agreed to a truce, and by April a ceasefire was extended until 1525. Once again Copernicus was engaged with the peasant farmers, this time resettling farms that had been abandoned on account of the fighting. But he had hardly begun this task when he was called back to Frauenburg. He was assigned as commissioner to help restore the shattered government, and by October he had also acquired a replacement for his destroyed house.

What was left of the war and its truce collapsed in 1525 when the nephew of the Polish king and grand master of the knights, Duke Albrecht of Brandenburg, converted to the Lutheran religion and dissolved the Teutonic Order. This left the Protestant duchy of Prussia as a fief within the boundaries of Catholic Poland. In later years Copernicus was called to give medical assistance to the Prussian court in nearby Königsburg (Kaliningrad).

Meanwhile Varmia and the Prussian surroundings were having problems of another sort, the near collapse of their currency (which was presumably on a par, but with a separate coinage, for each country). The coinage was based on a soft metal, silver, with an admixture of copper to strengthen the coins. While administering the farmlands from Olstyn, Copernicus became aware of the fact that the neighboring Teutonic Knights had increased the percentage of copper, thereby degrading the value of silver in the coins. Those in the know (which did not include the peasants) used the debased Teutonic coins to buy the older coins, which could be melted down for the silver, thereby causing a shortage of coins. In 1513 Copernicus wrote a brief Latin essay explaining how bad currency drives out the good, a rule now called Gresham's law, though Sir Thomas Gresham lived a generation later. Alas, Copernicus never published his analysis.

In 1519 Copernicus prepared a German version of his essay on currency, which should have ensured greater readership, but this was just at the time the conflict with the Teutonic Knights was heating up. The matter simmered along for nearly a decade, and then Copernicus was sent to present his ideas at the regional legislative assembly. We know of his lucid and far-ranging economic ideas only because of the survival of various manuscripts, and not because of their adoption at that time.

Near the end of the decade, on March 12, 1529, the fifty-six-year-old Copernicus observed the occultation of Venus by the moon, the last observation that entered into his astronomical treatise. He nevertheless continued observing, recording data about eclipses in his copy of Stoeffler's *Calendarium Romanum magnum*. (Incidentally, it was a pair of hairs in the gutter of this very volume in the Uppsala University Library that provided the DNA which served identify his bones a few years after his grave in the Frauenburg Cathedral was rediscovered in 2005.)

While the manuscript grew fatter, Copernicus made no preparations to have his book published. It would be a complicated book with many tables and diagrams, and there was no suitable printer in the entire region. He was getting older, still a busy senior member of the Cathedral Chapter, but it looked increasingly likely that his masterpiece would simply be destined for a dusty shelf in a sleepy cathedral library.

Chapter 6
Rheticus

As Copernicus entered his sixties, he was still a busy canon, particularly with respect to his medical expertise. From afar in neighboring Prussia, Duke Albrecht sought his advice for a sick friend, so Doctor Copernicus made the journey to Königsburg. When Bishop Maurice Ferber fell ill, Copernicus journeyed to set him on the path to health, and he went again when Ferber suffered a stroke; but two years later, when another stroke occurred, Copernicus was too late.

Meanwhile, Luther's Reformation was gaining power, and Ferber must have felt compelled to keep his own Catholic organization on the straight and narrow path. The majority of the Frauenburg canons were not ordained priests, so there was often a shortage of persons available to conduct the mass. Ferber pressed his canons to consider ordination. Furthermore, a few of the canons kept housekeepers, and Alexander Scultetus, one of Copernicus's friends, had several children. Ferber very diplomatically suggested that Copernicus's sleep-in housekeeper should go. After Ferber's death, Johannes Dantiscus, who had been angling for the Varmian bishopric himself and who had tried and failed three previous times, finally got the coveted position. Having sired an illegitimate son and daughter while a diplomat in Spain was apparently not a serious determent.

Dantiscus had known of Copernicus's cosmological work for many years and apparently admired it. Nevertheless, he warned Copernicus about his housekeeper, who was a distant relative. He also went after Scultetus with particular ferocity, driving him out of Varmia. And Dantiscus issued a death threat to any Lutherans caught in the Varmian territory. It is against this background that the final dramatic episode in the Copernican story played out.

The prologue begins 350 miles to the west, in Wittenberg, home of Martin Luther and his thriving university. The faculty included two mathematician-astronomers, the slightly senior Erasmus Reinhold, who taught astronomy, and Georg Joachim Rheticus, who taught algebra and trigonometry. "Rheticus" was a toponym, indicating that Joachim had come from Rhaetia in southeastern Austria. His birth name was Iserin, but when he was ten years old his father was convicted as a swindler and thief, and beheaded. The traumatized youngster was obliged to abandon his birth name, so when he matriculated at Wittenberg in 1532 he signed in under his mother's maiden name, Georgius Joachimus de Porris, though everyone called him Joachim.

Philipp Melanchthon, Luther's education lieutenant, promptly recognized Joachim's talent for mathematics, and they both shared a fascination with the big-picture elements of the universe, particularly as they might reveal themselves via astrology. But Joachim fell in with a rebellious satirical poet, and by 1538 there was good reason for him to make himself scarce for a while. He traveled to Nuremberg to see Dr. Schöner, one of the leading astronomical authors in those days. Joachim was eager to learn what was new in astronomy, and it was probably from Schöner that he heard of the reports about a radical new cosmology in Poland. Fired with curiosity and enthusiasm, young Rheticus (as he was now becoming known) resolved to travel unannounced to Varmia, dangerous as this might be to an astronomer from the very heart of the Lutheran reformation.

Together with a young colleague, Rheticus left Wittenberg early in May 1539 and reached their halfway point, Posen (Poznań), on May 14. From here Rheticus sent a progress report to Schöner in Nuremberg. Then, after passing through Copernicus's birth town (Torun), the young men reached Frauenburg late in May. We can only speculate on how they made the journey. Possibly they used the sixteenth-century equivalent of a rental car, buying horses and selling them near the end of the journey. The final major segment from Torun north to Malbork could well have been made by boat on the Vistula river.

Copernicus must have been both surprised and flattered by the unanticipated arrival of a young mathematics teacher from Wittenberg. For decades he had labored on his great project without students or serious understanding of his labors by his fellow canons. Here, at last, was a well-trained, twenty-five-year-old scholar eager to learn the arguments for the radical heliocentric cosmology.

And not only that, Rheticus had brought a gift: three handsomely bound volumes of considerable interest to Copernicus. First was the recently published Greek edition of Ptolemy's *Almagest*. Second was a pair of books printed by Johannes Petreius in Nuremberg, a treatise by Petrus Apianus and another by the thirteenth-century Polish optician Witelo. And the third volume contained another imprint from Petreius's press, Regiomontanus's work on triangles, together with the first Greek edition of Euclid's *Elements*. The three books from Petreius's press were subtle demonstrations of his ability to publish technical materials in a handsome form, something then impossible in Poland.

Eventually Rheticus helped get something else of considerable value to Copernicus, some precious observations of the planet Mercury, which he could have requested from Schöner, whom he had visited in Nuremberg, as mentioned earlier. Copernicus had intended to include four modern positions of each planet in order

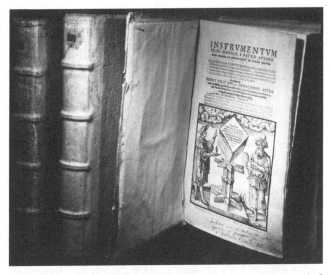

6. Rheticus brought these handsomely bound volumes of astronomical and mathematical texts as a gift for Copernicus. The inscription to his teacher is visible at the bottom of the title page.

to check whether the basic orbital constants had changed since the time of Ptolemy. This was easy for the three outer planets, Mars, Jupiter, and Saturn. However, the slow changes in the orientation of Venus's orbit completely defeated him, for the required geometrical configurations no longer occurred. Thus he simply had to assume that Ptolemy's ancient observations of Venus were good enough. For Mercury the situation was different. Mercury is always seen close to the sun, and at certain critical points in Mercury's path the planet was too low in the sky to be properly observed. This had deceived Ptolemy, who as a result made the geometry for Mercury unnecessarily complicated. In his *De revolutionibus* Copernicus mourned the great mists over the Vistula and declared that the tilt of the ecliptic at his more northerly latitude "rarely allowed us to see Mercury. Even at its greatest distance from the sun it does not rise within our view if it is in Aries or Pisces, nor do we see it set if it is in Virgo or Libra.

Consequently this planet has tormented us with much toil and many roundabout ways in trying to investigate its wanderings." Some commentators have concluded that Copernicus never saw Mercury, but that is unlikely; it's just that he could never sight it when it was at certain critical points in its path.

Clearly, when Rheticus arrived in Frauenburg, Copernicus's magnum opus was not yet ready for the printer. Among other things, the new positions for Mercury meant an overhaul for a few chapters near the end of the book. With Rheticus's encouragement and possibly direct help, Copernicus set to work to finish calculating the tables that had been revised as the basic numbers were adjusted. These tables made it possible to calculate the positions of the planets for any time past or future. It was a magnificent accomplishment, computing all the tables and showing how it was done. For Rheticus it was a fabulous tutorial, and he was ever keener to carry a copy of the precious manuscript to a German printer.

Still Copernicus hesitated. He confided to Rheticus that he would be very pleased if his geometry could predict the positions of the planets to within a sixth of a degree. Indeed, he would be as joyful as Pythagoras was in discovering his famous theorem. But he must have realized that he had not met this mark. Perhaps he could just prepare an almanac, without revealing the heliocentric cosmology behind it. He surely knew in his heart that the heliocentric arrangement would not by itself make the predictions more accurate except for his care in updating the orbital constants. Above all, he didn't want to be hooted off the stage.

Months passed as Rheticus learned more details of the book's structure, and as he helped tidy up the calculations. Surely Dantiscus knew there was a Lutheran visitor involved, and Copernicus was no doubt apprehensive at times. In the summer of 1539 Copernicus and his eager young scholar took up an invitation to go the safer neighboring province where his friend Tiedemann

Giese was bishop. It was there that Rheticus got the idea to write out a "first report," which he did after they returned to Frauenburg. By September 23 he had finished it, and after receiving the approval of both Copernicus and Giese, he headed to Danzig in search of a printer.

Not until page 19 did his account break the news that it was describing a *heliocentric* universe with a moving earth. But then he opened with full ammunition, albeit a technical description aimed at astronomers, including the dedicatee, Johannes Schöner in Nuremberg.

"Nature does nothing without a purpose," he wrote, quoting from the ancient physician Galen, and he proceeded to argue from the economy of nature: "Since we see that this one motion of the earth satisfies an almost infinite number of appearances, should we not attribute to God, the creator of nature, that skill which we observe in the common makers of clocks?"

It was a polemic, and at times turgid, report. Yet there were moments when the text soared:

> But if anyone desires to look either to the principal end of astronomy and the order and harmony of the system of the spheres or to ease and elegance and a complete explanation of the causes of the phenomena, by the assumption of no other hypotheses will he demonstrate the apparent motions of the remaining planets more neatly and correctly. *For all these phenomena appear to be linked most nobly together, as by a golden chain,* and each of the planets, by its position and order and every inequality of its motion, bears witness that the earth moves and that we who dwell upon the globe of the earth, instead of accepting its changes of position, believe that the planets wander in all sorts of motions of their own.

The scheme hatched by Rheticus to publish a *First Report* (*Narratio prima,* 1540) achieved his desired end. By 1541 a reprint was issued in Basel. There was no reaction to call Copernicus a heretic; nor was he hooted off the stage. He raced to finish the loose ends and to have his manuscript copied so that Rheticus could carry it back to Germany in search of a publisher, presumably Petreius in Nuremberg, who had already expressed an interest in the book.

Armed with the manuscript of Copernicus's magnum opus, early in the autumn of 1541 Rheticus bid his teacher farewell and headed back to Wittenberg. In retrospect, his two-year-plus sojourn at Frauenburg proved to be a crucial link in bringing *De revolutionibus* and the heliocentric cosmology onto the world stage.

Chapter 7
De revolutionibus

Soon after Rheticus returned to Wittenberg in the fall of 1541, he was elected dean for the first semester of the term. This was not so much an honor as an academic chore; hardly any of his colleagues served a following term. Rheticus hoped not to be locked in too tightly at Wittenberg, for he had the precious manuscript of Copernicus's book, waiting for the printing to be arranged in Nuremberg.

Rheticus had thought ahead as he was leaving Frauenburg. He had met Duke Albrecht in Königsburg when Copernicus had taken him there on a three-week medical mission, and he knew that the duke had connections in Wittenberg. So he wrote a letter explaining how he needed another leave to go to Nuremberg to oversee the printing of *De revolutionibus*. Whether the response from one of the duke's assistants played a serious role in freeing Rheticus after a year of teaching we can't be sure, but the favorable outcome was surely satisfying to Rheticus and pleasing to Copernicus as well.

Meanwhile Rheticus took the mathematics section of Copernicus's book to a local printer in Wittenberg, so that his students would have access to a table of sines, one of the first such tables ever printed. This was part of Copernicus's modernizations of Ptolemy's mathematics. Ptolemy had used an earlier and clumsier trigonometric arrangement, the so-called chord function. Sines and cosines had been introduced into astronomical calculations during the golden age of Islamic mathematics (by al-Khwarizmi in the ninth century). Rheticus himself would become one of the great calculators of trigonometric tables, and in the Wittenberg version of the Copernican tables he increased the number of digits that his teacher had provided.

By early spring of 1542 Rheticus was busy at Petreius's print shop in Nuremberg as the printing of *De revolutionibus* itself began. The text was printed on pairs of sheets twice as wide as a single page. In the morning one side of the sheets (which had been slightly dampened the previous evening to produce a better impression of the type) was printed, and in the afternoon before the paper itself could dry, the back sides of the sheets were printed. When dry, the sheets were folded and the pairs inserted one inside the other. This gathering of eight pages was called a signature.

Individual chapters began with a special large-size initial, and for this purpose Petreius had a single decorative alphabet, which he had been using for many years on his better publications, including a Latin Bible that he had printed in 1527. The first signature of *De revolutionibus* included six short chapters, the second included four chapters, and the third two more chapters, but by chance all three of the first signatures had a chapter beginning with the letter Q. Because Petreius had only a single set of the special initial alphabet letters, he had to break down the first set of forms to retrieve his Q before his assistants could finish setting the type for the second signature. This meant that a professional copy reader—Rheticus or his substitute—had to be on site almost continuously for the production of the book to proceed on schedule.

Copernicus's volume was divided into six books, as the full title explained: *De revolutionibus orbium caelestium libri sex*, that is, *Six Books on the Revolutions of the Heavenly Spheres*. Each book was in turn divided into chapters. The manuscript opened with a brief essay in praise of astronomy and the beauty of the heavens, "established by the best arrangement and directed by divine management." Because Copernicus had subsequently written a new preface, the original essay was scrapped, and the first book began with a series of short chapters introducing the general cosmological background, such as "The movement of the celestial bodies is regular and circular, everlasting—or else is composed of circular parts."

The initial chapters led up to chapter 10, a soaring cosmological accolade. As this one unfolded, Copernicus introduced the most compelling aesthetic argument for the heliocentric arrangement. He laid out the order of the planets beginning with Saturn, which completes its circuit in thirty years, then Jupiter, which moves in a twelve-year period of revolution, followed by Mars, which completes its round in about two years. Next comes the earth with the moon, then Venus with a period of nine months and finally Mercury in eighty days. (Here Copernicus was speaking with poetic license, since he knew full well that the Venusian period was 7½ months and Mercury's was eighty-eight days.) "In the center of all rests the sun," Copernicus declared, "for who would place this lamp of a very beautiful temple in a better position than this wherefrom it can illuminate everything at the same time?" "In this ordering we find that the universe has a wonderful common measure [the size of the earth's orbit] and a sure bond of harmony for the motion and size of the orbital circles such as cannot be found in any other way."

He turned next to one of the most powerful geometric arguments in favor of the heliocentric system, the retrograde motion of the major planets. In the Ptolemaic system these planets always went into their backward motion when they were opposite the sun in the sky. Precisely why this happened was totally unexplained. This is what the medieval logicians called a *quia* argument, a "fact in itself." With the sun-centered arrangement the retrogression opposite the sun became a "reasoned fact," for the first time an explanation of the cause of this puzzling phenomenon. However, no such phenomenon was seen with respect to the fixed stars. This, Copernicus realized, was because of their immense distances. "So vast, without any question, is the divine handiwork of the almighty creator."

What had originally been planned as Book 2, with a short section on geometrical theorems and the new table of sines, was now added to complete the first of the six books. Then, perhaps while the craftsmen began preparing the first of the 142 woodblocks required to illustrate the geometry of the book, Rheticus took off a few weeks for a family visit around Feldkirk in southeastern Austria.

By the time he returned to Nuremberg, a new professorship awaited him in Leipzig, with a special salary of 140 florins, well above the normal salary of 100 florins. For the university in Leipzig, Rheticus was a major "catch" from Wittenberg, the leading Lutheran university, and he realized that it was a tough choice between his supervising the printing of *De revolutionibus* and accepting the prestigious post in Leipzig. Reluctantly, he turned over his proofreading task to Andreas Osiander, a local clergyman well trained not only in Latin and Greek but also in mathematics. And under Osiander's sharp eye, the printing continued.

We do not know precisely where in the printing of the book the transition took place, but possibly somewhere in Book 2. While *De revolutionibus* tends to follow the structure of Ptolemy's *Almagest,* one of the most significant differences occurs with Book 2. Here Copernicus placed his star catalog, 1,024 stars arranged in forty-two constellations. It was not an original compilation, but something adapted from one of various lists derivative from the *Almagest*. But it was a list with a difference.

It is comparatively easy to make a star map showing the relative positions of the stars. The north celestial pole provides a unique point for determining how far north or south a star lies. Likewise for geographical coordinates, it is in principle easy to establish the latitude of a city or an island by measuring the altitude of the north celestial pole. But to measure longitude requires adopting a fixed starting point, so the British could measure with respect to London, the French with respect to Paris, and so on. For astronomers the choice of an east-west reference point was easier because there was the intersection between the celestial equator and the ecliptic path of the sun (the great circle that runs through the zodiac, tilted about 23° with respect to the equator). This is not necessarily an easy choice, however, because the equatorial circle can be traced in the starry night and the ecliptic in the sunny daytime, and it takes real ingenuity to discover where the intersection actually is. Not only that, the intersection point is slowly but constantly moving with respect to the stars.

Suffice it to say, Copernicus disagreed with Ptolemy about which of these two circles was the more fundamental one. According to Ptolemy, above the solidly fixed earth the entire sky was twirling about its axis every day. In contrast, in the Copernican system the sky was solidly fixed and the earth was spinning on its axis, while the orientation of the axis was slowly precessing just as the axis of a toy top does. Thus, Copernicus picked not an invisible point as the starting position for his star catalogue, but an actual star in the constellation Aries. For him, the starry frame remained unchanging and could therefore come first in his volume. Ptolemy, on the other hand, needed to place his star catalog farther back in his book, after he had discussed the motion of the sun and moon; these were used to determine where the invisible ecliptic began at any specified time. Copernicus made his point about the solidly fixed heavens for pedagogic reasons, but no other work would follow his idiosyncratic lead.

In Books 3 and 4 he addressed the paths of the sun and moon. The precession of the equinoxes since the time of Ptolemy was a new issue for bringing astronomy up to date. Likewise the slow drift in the celestial position of where the sun was closest to the earth (the so-called perihelion) needed updating. As for the moon, there was an interesting correction to be made. Ptolemy had already determined that the moon moved in an eccentric orbit around the earth, but the eccentricity was effectively varying in a thirty-two-day cycle. He added an extra circle controlling the moon's position to account for this "evection." With this addition the longitudinal errors of prediction were kept under a degree, but the model radically erred in calculating the moon's distance from the earth and hence its apparent size in the sky. Ptolemy passed over this anomaly in silence, but by the golden age of Islamic astronomy, that model received sharp criticism because sometimes the moon was calculated to be twice as large as usual. In his own treatment, Copernicus changed the order of the extra circle and greatly reduced the predicted changes in the apparent size of the moon, but otherwise his predicted lunar longitudes were about as unsatisfactory as Ptolemy's.

Book 5, by far the longest in *De revolutionibus*, dealt with the planets, beginning with the longitudes of Saturn and continuing to the most difficult case, Mercury. As noted earlier, Copernicus had difficulty getting the observations he needed for Mercury, and when fresh data finally arrived from Schöner in Nuremberg (quite probably with help from Rheticus), this section must have been among the last to be completed. Included were "mean motion" tables to find an approximate place for each planet, and then there were "equations," tables to correct for the position of the earth as well as for the variable speed of each planet as it rounded the sun. It took a fair amount of trigonometry to set up the equation tables, but once the tables were at hand most of the computation for a predicted position was completed simply by addition and subtraction, with occasional multiplication. (An example is given in our second appendix.)

The short and final Book 6 was on planetary latitudes, that is, the calculation of a planet's distance north or south of the ecliptic path of the sun. For better or worse (and it was definitely for worse), both Ptolemy and Copernicus resorted to a separate geometrical mechanism for calculating the planetary latitudes compared to the longitudes. Neither Ptolemy nor Copernicus specified any observations on which their predicted latitudes were based. The numbers they used for the tilts of planetary paths were apparently traditional values, possibly going back to the Babylonians. Several generations later, the astronomer Johannes Kepler (who was born ninety-eight years after Copernicus) declared that astronomical models had to be physically real, and using one scheme for finding longitudes and another one for latitudes was unacceptable. What we think of as the Copernican system today was actually made workable by Kepler.

Meanwhile, as the first signatures were printed, sets of pages were sent to Copernicus, who carefully proofread them. Some of the errors were already in his original manuscript, where they had escaped detection. Although the errors on the pages already printed couldn't be corrected, Copernicus did send the list of errata back to Nuremberg, and in the end Petreius printed an errata sheet.

Chapter 8
The book nobody read

Frauenburg, Thursday, May 24, 1543: the mail courier from Nuremberg came with a long-awaited package. It included the final section to be printed of *The Revolutions of the Heavenly Spheres*, that is, the front matter of the book including the title page, the author's preface and dedication to Pope Paul III, and the detailed list of chapter titles.

There was something else lurking there, but Copernicus probably didn't notice. Some months before he had suffered a cerebral hemorrhage, the stroke paralyzing the right side of his aging body. As Georg Donner, his fellow canon, reported to Bishop Tiedemann Giese, "his memory and mental powers had abandoned him."

But both Giese and Rheticus noticed something that outraged them. On the opening immediately following the title page was an anonymous *Ad Lectorem*, "To the Reader Concerning the Hypotheses of this Work." This essay was not part of Copernicus's manuscript, and it was an addition not authorized by Rheticus. Despite Giese's threatened lawsuit against the printer Petreius, nothing was done about this unauthorized and anonymous insertion. In part this introduction read:

> Now since the novelty of the hypotheses set forth in this work have become widely known, I have no doubt that some scholars will have

58

taken great offense and think that it is wrong to churn up the rightly time-honored liberal arts. Yet, if they judge the matter carefully, they will find that the author of this work has done nothing blameworthy. For it is the duty of an astronomer to record the motions of the heavens with diligent and skillful observations, and then he has to propose their causes. Or rather, hypotheses, since he cannot hope to attain the true reasons.

Our author has done both of these very well. For it is not necessary for the hypotheses to be true, nor even probable; it is sufficient if the calculations agree with the observations. . . .

Therefore, we must allow these new hypotheses to take their place among the old ones—which are no more probable—especially since they are admirable and easy, and since they bring with them a vast treasury of very learned observations. But, as far as hypotheses are concerned, let no one expect anything certain from astronomy, which cannot provide it, lest he take as true something constructed for another purpose, and leave this discipline a greater fool than when he entered.

Gradually a few people learned that this essay did not have Copernicus's permission. The reaction by Kepler's teacher, Michael Maestlin, is particularly interesting, a stirring blast at this unauthorized addition. Above the *Ad Lectorem* in his copy of the book, Maestlin wrote:

This preface was added by someone, whoever its author may be (for indeed, its weakness of style and choice of words reveal that it is not by Copernicus), lest someone at the mention of these hypotheses would hiss them off the stage as false and unworthy of reading, or would approve them at first glance injudiciously out of love of novelty; rather he ought first to read them, reread them, and only then judge them. However, once convinced, he will then afterward be unable to oppose them. But the author of this letter, whoever he is, while he wishes to ask the reader neither to reject nor accept these hypotheses casually, yet he chatters away foolishly on some things on which it would have been better for him to have remained silent. *For the sciences are not strengthened by shaking their foundations.* But since the very material defends itself so well, he [the anonymous author] labors here in vain, seeing that he is wanting in strength and arguments. And therefore, I simply cannot approve, much less defend, any of the principal points of this letter.

Subsequently Maestlin added another short paragraph on the facing page:

> Among the books of Philipp Apian (which I bought from his widow) I found the following words, which I conjecture he had copied from somewhere: On account of this letter Georg Joachim Rheticus, the Leipzig professor and disciple of Copernicus, became embroiled in a very bitter wrangle with the printer, who asserted that it had been turned over to him with the rest of the work. Rheticus, however, suspected that Osiander had prefaced it to the work; if he knew this for certain, he declared, he would treat the fellow in such a manner that in the future he would mind his own business and not dare to aggravate astronomers any more. Nevertheless, [Peter, Philipp's father] Apian told me that Osiander had openly admitted to him that he had added this all by himself.

Maestlin's very short third and last note has been squeezed into the remaining space over the *Ad Lectorem:* "NB. I know for sure that the author of this letter was Andreas Osiander."

How did Maestlin finally know for sure that Osiander, the proofreader, had written the warning to the reader? There may have been many ways he could have found out, but one very certain way was that his young student, Johannes Kepler, owned a second-hand *De revolutionibus,* in which the previous owner, Jerome Schreiber, had written Osiander's name above the *Ad Lectorem.* Schreiber was an insider who received his copy of the book directly from the printer, and who had entered Wittenberg University in the same class as Rheticus. We know that Kepler

NICOLAI CO

PERNICI TORINENSIS

DE REVOLVTIONIBVS ORBI-
um coeleſtium, Libri $\overline{VI.}$

Habes in hoc opere iam recens nato, & ædito,
ſtudioſe lector, Motus ſtellarum , tam fixarum,
quàm erraticarum, cum ex ueteribus, tum etiam
ex recentibus obſeruationibus reſtitutos: & no-
uis inſuper ac admirabilibus hypotheſibus or-
natos. Habes etiam Tabulas expeditiſsimas , ex
quibus eoſdem ad quoduis tempus quàm facilli
me calculare poteris. Igitur eme, lege, fruere.

Ἀγεωμέτρητος οὐδεὶς εἰσίτω.

Ioachimus Rheticus presbyter Lipsiæ
suo Andreæ Aurifabro d.d.
20 Aprilis anno 1547.

Norimbergæ apud Ioh. Petreium,
Anno M. D. XLIII.

7. On this title page of *De revolutionibus*, printed in Nuremberg in
1543, Rheticus has written an inscription for Andreas Aurifaber, dean
of the University of Wittenberg. The partially crossed-out words
"orbium coelestium," meaning "of the heavenly spheres," were added
by the printer, although Copernicus preferred the shorter title.

showed his book to his teacher, because on one critical page Maestlin himself added an annotation.

Eventually Kepler revealed to the world that Osiander was the author of the anonymous letter to the reader; he placed an announcement on the back side of the title page of his greatest book, the *Astronomia nova*. This book, subtitled "coelestial physics," was indeed the "new astronomy," adding physical principles to support Copernicus's heliocentric premise. Unlike his illustrious predecessors, he demanded a physical model that worked for both the longitude and the latitude of the planets. What passes today as the "Copernican System" is in detail the Keplerian system.

Despite the immense amount of ink spilled on Osiander's intrusive interpretation of Copernicus's intentions, it really did protect the book against traditionalists, who were persuaded that *De revolutionibus* was indeed a safe recipe book for computing the positions of planets, but not to be considered as actual physical reality. And the traditionalists included not only theologians, philosophers, and churchmen but astronomers as well.

Among the twentieth-century writers who "didn't get it" was the distinguished novelist Arthur Koestler, who apparently thought the heliocentric system was so obvious that it should have been quickly adopted, and if it didn't receive the rapid approval it deserved, then almost no one was paying attention. Consequently, in his best-selling history of astronomy, *The Sleepwalkers,* he described *De revolutionibus* as "the book nobody read" and "an all-time worst seller."

In 1970, as we were approaching the quinquecentennial of Copernicus's birth, I stumbled onto a discovery that would change my career from astrophysicist to historian of astronomy. I was aware of Koestler's claim that nobody in the sixteenth century was reading Copernicus's technical tome when, quite unexpectedly, at the Royal Observatory of Edinburgh I found a first edition *De revolutionibus* whose margins were filled with Latin annotations from beginning to end. Clearly here were the traces of a reader who was paying attention. Except on the title page he didn't write, "this book stops the sun and throws the earth into swift motion." Instead, he paraphrased the title of Book 1, chapter 6: "Celestial motion is uniform and circular or composed of uniform and circular parts." Here was a view of *De revolutionibus* as a recipe book par excellence! But whose copy was it?

With a little lucky sleuthing, I found that the annotations were in the hand of Rheticus's senior colleague at Wittenberg, Erasmus Reinhold, the leading astronomical pedagogue in the generation following Copernicus. Reinhold had almost nothing to say about the radical cosmology in Book 1, but in the technical chapters that followed he was full of commentary. He recognized that some of Copernicus's calculations could be modestly improved, but it apparently never occurred to him that the accuracy of the observations didn't warrant refinement. He eventually worked up Copernicus's data into the perfect recipe book, the so-called *Prutenic Tables*, which made calculating almanacs and ephemerides very straightforward—and the process was independent as to whether the reader considered the earth or the sun as the fixed reference frame.

Finding such an annotated book inspired me to look for more copies, mostly with the hope that I would find something interesting to say when Copernicus's five hundredth anniversary rolled around in 1973. Such was the beginning of the search to examine every surviving copy of the first, 1543 edition and also the second, 1566 edition. This took not three years but thirty, with the inspection of books from Berkeley to Budapest, from Guadalajara to Glasgow, from Moscow to Melbourne, altogether nearly six hundred copies.

One of my first major findings was that several of Reinhold's advanced students transcribed his annotations into their own copies, and in turn their own students made further copies. It was difficult to find any major annotations that didn't appear in other copies. It was as if there were invisible colleges of students preserving the remarks of charismatic teachers. Simultaneously *De revolutionibus* became a famous book, and the libraries of the nobility often included it—except they rarely annotated their copies. Although the heavily annotated copies were fascinating they were comparatively rare.

In 1973, the Copernican anniversary year, I discovered in the Vatican Library another well annotated *De revolutionibus* derivative from Reinhold's copy, but with an intriguing set of diagrams that led to an image very close to the system proposed in 1588 by the Danish astronomer Tycho Brahe. In Brahe's cosmology the earth rather than the sun was at the center. The sun cycled around the fixed earth, carrying the other planets as they cycled around the moving sun. Tycho's was a compromise system, much admired by the Jesuits despite the fact that Tycho was a Lutheran. I soon discovered that the handwriting in the Vatican volume matched a copy in Prague, long believed to be Tycho's personal working copy. I was excited to think that I had found the working copy where the Danish astronomer was just developing his alternative cosmology.

There is a well-known aphorism in academia, "Publish or perish!" Those Copernican volumes at the Vatican and in Prague were not, in fact, annotated by Tycho Brahe, but by a man who never published and whose reputation almost perished. An itinerant scholar from Wroclaw (Wratislavia), Paul Wittich owned and annotated at least four copies of *De revolutionibus*, and in his travels visited Tycho in Denmark and showed him some of his planetary diagrams. Later, after Wittich's early death, Tycho waged a lengthy campaign until he finally captured Wittich's annotated copies. This is not to say that Tycho plagiarized

Wittich's alternative cosmology, but it is part of the back story on cosmological movements of the sixteenth century. And uncovering the relationship of those four annotated Wittich volumes was part of the long bibliographic archaeology of the Copernican texts.

An early mention of Copernicus's cosmology appeared in print in 1556, in Robert Recorde's popular dialogue *The Castle of Knowledge*. The Master cites Copernicus, "a man of greate learninge, of much experience, and of wonderfull diligence in observation," to which the young Scholar exclaims, "Nay syr in good faith, I desire not to heare such vaine phantasies, so farre against common reason, and repugnante to the consente of all the learned multitude of Wryters, and therefore let it passe for ever, and a daye longer."

To which the Master replies, "You are too younge to be a good judge in so great a matter: it passeth farre your learninge and theirs also that are muche better learned then you, to improve his supposition by good argumentes, and therefore you were best to condemne no thing that you do not well understand."

A generation later, in 1576, a young English mathematician, Thomas Digges, published the first translation of a part of Book 1 of *De revolutionibus* under the title *A Perfit Description of the Coelestiall Orbes*. This new edition of his father's perpetual almanac was the first of nine editions with the Copernican section.

The aftermath of heliocentric cosmology

Digges stated that he had done this so that "noble English minds might not be altogether defrauded of so noble a part of Philosophy." His presentation included a diagram of the Copernican system, but with an extraordinary novelty: there was no longer a starry spherical shell, but a field of stars extending ever outwards. The caption declared, "This orbe of starres fixed infinitely extendeth hit self in altitude sphericallye, and therefore immovable; the pallace of foelicitye garnished with perpetuall shininge glorious lights innumerable, farr excellinge our sonne both in quantitye and qualitye; the very court of coelestiall angelles, devoid of greefe and replenished with perfite endlesse ioye, the habitacle for the elect."

Earlier celestial images had included God and the elect in a heaven just beyond the sphere of fixed stars, so Digges had devised an awesome solution for the changed place of heaven in the new cosmology. Incidentally, the great Copernicus chase turned up Digges's own copy at the University of Geneva in Switzerland. At the top of the title page he wrote his personal endorsement of Copernicus's book: *Vulgi opinio error* ("The opinion of the common folk is in error").

Not until 1596, more than fifty years following publication of *De revolutionibus,* did an enthusiastically Copernican treatise appear. This was Kepler's *Mysterium cosmographicum,* or *Sacred Mystery of the Cosmos.* This astonishing book attempted to account for the otherwise puzzling spacing of the planets in the Copernican system by inserting the five Platonic solids—octahedron, icosahedron, dodecahedron, tetrahedron, and cube—between the six planetary spheres. For example, a cube just fit between the spheres of Saturn and Jupiter., and a tetrahedron between the spheres of Jupiter and Mars.

A friend of Kepler's was traveling to Italy, so he sent along a couple copies of his little book, destined for whoever might appreciate them. Near the end of the trip both copies were given to Galileo, who dashed off a quick thank-you note and mentioned that he was also a Copernican, but privately. Kepler, who may never have heard of Galileo, responded, urging him to stand forth as a Copernican, but for a decade Galileo maintained his silence.

In 1609 Kepler published his most memorable book, *Astronomia nova.* It truly was the *New Astronomy, based on causes, or celestial physics.* It was not the physics that finally held, but his insistence that physical causes were essential, which transformed the requirements for a satisfactory theory. It basically established what is today called the Copernican system.

Meanwhile, in 1609 Galileo was converting a new carnival toy into a scientific research instrument, the telescope. In January 1610 he discovered the little starlike companions of Jupiter and on the night of the thirteenth was himself converted from a reluctant Copernican into a crusading heliocentrist. He had that night realized that those little stars were satellites moving around Jupiter, a veritable miniature Copernican system. His *Sidereus nuncius* quickly followed, where he announced the earthlike features on the moon, the Jovian satellites, and the host of faint stars not seen by the naked eye.

Subsequently Galileo began to campaign for the Catholic Church to maintain an open position with respect to the heliocentric cosmology. But the Church, having been driven into a more literalistic position as a consequence of the Protestant Reformation, took a conservative interpretation of Scriptures, including Psalm 104, which stated "The Lord my God laid the foundations of the earth that it should not be moved forever."

As a consequence of Galileo's campaign, the Vatican decided to place *De revolutionibus* on the *Index of Prohibited Books*. The Roman index placed many books on its list "until corrected," but in the 1664 edition of the *Index*—the first to include *De revolutionibus* as well as Galileo's *Dialogo*—Copernicus's book was the only one for which a detailed list of corrections was specified. Most of the ten changes were designed to make the contents simply hypothetical and not a description of physical reality. Because the corrections are so specific, it can be quickly established whether or not a copy was censored. Approximately two-thirds of the copies in Italy were censored, but virtually none in Catholic countries such as France or Spain. Apparently the rest of the world looked on the exercise as a local Italian imbroglio and were having none of it. In fact, the Spanish version of the *Index* explicitly permitted the book.

Eventually Galileo wrote the book that won the war. His *Dialogue on the Two Great World Systems* (1632) made belief in a moving earth intellectually respectable. But for Galileo the *Dialogo* was a personal disaster. He had recklessly placed the pope's favorite argument against the motion of the earth into the mouth of the character Simplico, which the Italian readers recognized as a pun on "simpleton." Pope Urban VIII, who was already walking a political tightrope between the French and Spanish cardinals, was not prepared to be forgiving, and Galileo remained under house arrest for the remainder of his life. Not until 1835 would the Catholic ban on Copernicus's book be lifted.

The search for copies of Copernicus's book turned up many interesting exemplars, including Kepler's and Galileo's copies—the latter was lightly censored—and the family of annotations stemming from Erasmus Reinhold. But where was Rheticus?

As part of my search for copies of *De revolutionibus*, I arranged an expedition to a series of provincial libraries in Italy, including the Biblioteca Palatina in Parma. Unfortunately the Palatina's first edition was nowhere to be found. But sometime later, when I learned that the book had long been missing, I asked for a copy of the entry in the library's card catalog. When the description arrived, I saw that the missing book was prefaced by a Greek poem in the hand of Joachim Camerarius, a classicist who had been Rheticus's colleague at the university in Leipzig. I knew of one book, undoubtedly the most fabulous copy in private hands, that matched this description. It had turned up, mysteriously, in the London book market shortly after World War II. Could that copy have been "liberated" by an Allied soldier or a hungry librarian? But a very careful inspection of that copy showed no traces of any defaced library markings. Nevertheless, the description in the Palatina catalog seemed a perfect match.

Before I had the opportunity to commit myself in print, this detective story entered another chapter. An eminent French book dealer, Pierre Berès, informed me he had something of interest concerning Copernicus. When I visited him in Paris, he told me he had been sent anonymously a packet of photocopies of title pages of early books, including not only *De revolutionibus* but also two pages of a Greek poem. When I explained the situation to Berès, he quickly concluded that these photocopies were made from the missing Palatina copy, and he would have nothing further to do with it. With that, the trail went cold, leaving me with a troubling dilemma: whether to publish the photocopies in my *Census* to alert possible buyers to a stolen copy, or to suppress this information lest the thief simply destroy the page with the Greek poem.

After a decade passed, and before I had to make a final decision, the book prefaced with Camerarius's manuscript poem surfaced once again. Word traveled quickly in the universe of book sellers, and presently I was telephoned by an expert at Sotheby's auction house in Milan. "We hear you are suspicious of our copy of Copernicus's book," the voice said. "What evidence do you have that the book belongs to the Palatina?"

I explained about the catalog description. Apparently I had had access to a special catalog, not the public one, for my information was news to Sotheby's representative. He nevertheless reported that most of the books being offered to Sotheby's, if not all, were listed in the Palatina catalog but were in fact missing from that library, strong circumstantial evidence for their true ownership. There the matter rested for several months. Then I received an e-mail from a book dealer in Florence stating simply, "Nicolaus is back in the Biblioteca Palatina."

I was delighted by this state of affairs, and unprepared for the next chapter. I had occasion to be in Padua, so I took the train over to Parma to check on the book with the Camerarius poem. But I was met by a group of long-faced librarians who explained that the book was actually in the hands of the police because the ownership of the book was under litigation. I was indignant because the ownership seemed beyond question. Since the Italian edition of my book was in its final editorial stages, I added a postscript to embarrass the authorities—a dangerous game, but one that achieved its goal. Within a few weeks of its publication I had an e-mail from the library, accompanied by images to show that they really had their book back.

And now for the denouement. I have rarely felt so welcomed to a library as when I returned to the Palatina. The book itself was a wreck, without a binding and so dirty that it had been washed. Fortunately the staff promptly provided an ultraviolet light, which made most of the faded annotations legible. On folio 96 I was stopped cold. There was an annotation that I knew stemmed from Rheticus. As I continued, I gradually realized that I had found Rheticus's own annotated copy. It was filled with many small corrections to the printed text, a set partly found by Copernicus, but probably many, particularly toward the end, from Rheticus himself.

The title page itself was entirely covered with notes. But it was illegibly smeared except for a single word, *hieronymo*. The final o made it dative case, meaning "to Jerome." My synapses were a little slow, but eventually its identity dawned on me. Soon after the book was printed Rheticus took a trip to Italy, visiting probably the greatest of sixteenth-century mathematicians, Jerome Cardano. This astonishing copy was very likely Rheticus's house gift to his Italian host.

Rheticus surely had another copy with the errors indicated, for these errors are marked in about a dozen other copies, partly his private legacy to his students.

Canon Copernicus's grave

The Frauenberg Cathedral has eight major columns on each side of the central aisle, each with an altar, and each of the sixteen canons was assigned to one of the altars. Copernicus's altar was the fourth on the right-hand side, and there was a tradition for each canon to be buried in the vicinity of his altar.

In 2004 Bishop Dr. Jacek Jegierski became troubled that there was no memorial of any kind for the most illustrious staff member the cathedral had ever had. Consequently he asked Jerzy Gąssowski, one of Poland's most distinguished archaeologists, for help in locating Copernicus's relics. At first Gąssowski declined on the grounds that there were far too many unidentified graves under the cathedral floor, so that, in his words, this would be "like looking for a needle in a haystack." But when he became convinced that the archival evidence made it likely that Copernicus would have been buried near the fourth altar on the right, he agreed to try.

In two summer seasons of excavation Gąssowski's team found thirteen burials, and the leading candidates for Copernicus's remains were scattered bones of a sixty- to seventy-year old man at the lowest level. The cranium was present, but not the mandible; the cranium was sent to the Central Forensic Laboratory of the Polish Police in Warsaw for a facial reconstruction. The results were favorable, but not entirely convincing. Gąssowski sought a DNA sample, for example from the astronomer's maternal uncle, Bishop Watzenrode, but that grave was not found, and collateral descendants could not be traced.

Then, when in a lecture Gąssowski gave in Sweden he expressed his hope of finding a DNA sample, an astronomer in the audience recalled that Copernicus's personal library had been captured and brought to Sweden in the Thirty Years War. Perhaps something in one of the books could yield a clue. Improbable as this sounds, nine hairs were found in the gutters of Copernicus's annotated copy of the *Calendarium Romanum magnum*. Four were well enough preserved for DNA enhancement to be undertaken. The mitochondrial DNA of two of the hairs matched each other as well as the DNA from a tooth in the cranium.

With great pomp and ceremony, on May 19, 2010, Copernicus's bones were again laid to rest within a meter of their original site, but this time below a handsome granite marker depicting the sun and its family of planets. Nicolaus Copernicus, canon of the Frauenburg Cathedral, had invented the solar system, and his *De revolutionibus* was the book that moved the world.

Appendix 1

Copernicus's alternative to the equant

We can imagine three major levels of sophistication in modeling the behavior of the planets in a heliocentric system. The first is the "kindergarten model," with the sun in the center, and the planets arrayed in circles centered on the sun, each with its own radius and period of revolution.

A second level of sophistication would allow each planet to have an off-centered or "eccentric" orbit, which means two additional fundamental parameters for each planet, to specify the degree of eccentricity and the direction on which the eccentric center lies. To take full advantage of this level of sophistication, it would be necessary to apply the same arrangement to the earth's orbit. This Copernicus did not do, because Ptolemy himself treated the earth as very special and adapted his calculation methods to this fact, which allowed substantial simplifications, some very advantageously made (and which reflected Ptolemy's mathematical genius).

It was Johannes Kepler, two generations later, who recognized this deficiency in Copernicus's heliocentric arrangement and who established the heliocentric system as we know it today. In seeking to find the physical laws governing the solar system, he stumbled onto the extremely subtle conclusion that the planets traveled in elliptical orbits around the sun.

Eventually a third level of sophistication emerged following the work of Isaac Newton. He started out supposing that the sun's gravitational attraction held the planets in their orbits, but his really brilliant discovery was that gravitation was *universal*: every planet (or satellite or star) attracted every other one, so the planetary paths were only approximately ellipses. Thus, to compute the orbits of the planets more exactly required taking all the planets' attractions into consideration, something undertaken by the more brilliant of Newton's successors.

Copernicus was the bold pioneer who took the major step into the heliocentric geometry. It was Kepler who remarked accurately but ungraciously that Copernicus interpreted Ptolemy rather than the sky. In this appendix we look at Copernicus wrestling with the second level of sophistication, to make the system not more accurate but more philosophically acceptable to the handful of his contemporaries who could appreciate it.

Copernicus stood at a crucial transition point as he revised the geocentric viewpoint, but he was unable to come to terms with the physical implications of the radical realignment. Within his own aesthetic vision he had to generate the nonuniform motions out of combinations of uniform circular motions, a requirement that went all the way back to Aristotle. Because this part of his cosmology turned out to be a dead end, modern secondary sources tend to ignore this rather complicated part of his *De revolutionibus*, even though it made up a bulky part of the book.

In order to account for both the nonuniform motion of the epicycle being carried around the deferent circle and also the varying size of the retrograde arcs, Ptolemy had introduced uniform motion about an empty spot called the equant, shown in the left half of Figure 8. Note that in the Copernican alternative, shown in the right-hand part of our diagram, the straight dashed line is always parallel to the line from the equant point. However, as noted below, the resulting dashed trajectory is not exactly a circle.

Ptolemy argued that his equant had been found in agreement with nature even though it was difficult to establish it from first principles. Ptolemy received heavy criticism for this by the Islamic astronomers, and Copernicus too expressed his dissatisfaction in his *Commentariolus*.

Copernicus originally chose to use a double epicyclet, precisely the same scheme suggested two centuries earlier by Ibn al-Shatir of

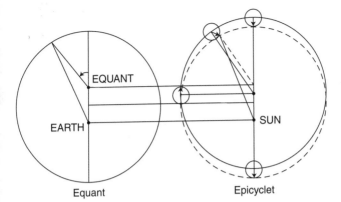

EQUANT

EARTH

SUN

Equant

Epicyclet

8. The diagrams show Copernicus's replacement of the Ptolemaic equant by a pair of uniform circular motions. The epicyclet always moves to form an isosceles trapezoid.

Damascus; by the time he prepared the *De revolutionibus*, he had simplified the scheme to the single-epicycle form shown here. The epicyclet rotates counterclockwise, making two revolutions as it traces out the figure shown by the dashed line. In general Copernicus credited the names of his predecessors, though he did not mention Ibn al-Shatir. Thus it is quite probable that the Polish astronomer had never heard of Ibn al-Shatir, though he could have picked up the Damascene configuration from some anonymous source. So far no convincing path of transmission has been traced. And of course this particular mechanism has nothing to do with heliocentric cosmology.

In *De revolutionibus*, Copernicus remarks that this "eccentrepicycle" will not exactly produce a circle. In the copy of Copernicus's book that Johannes Kepler inherited, on the facing page to that remark the previous owner had written *ellipse* in Greek. Of all the copies Kepler could possibly have inherited, it is remarkable that it was the copy containing this suggestive word. Kepler was astute enough, however, to realize that in Copernicus's case the figure bowed out where the ultimate configuration that Kepler found bowed in. In fact it was Kepler who ultimately showed that the search for strictly circular celestial motions was a dead end.

Appendix 2

De revolutionibus as a recipe book for planetary positions

In the middle of the title page of *De revolutionibus* there stands a small rectangular publisher's blurb. Translated, the Latin reads:

> Zealous reader, in this book just edited and published, you have the motions of the stars and planets restored from both ancient as well as recent observations, & moreover embellished by new and admirable hypotheses. You have also the most expeditious tables from which you can calculate their motions for any time whatsoever. Therefore buy, read, profit.

This was added clearly at the instigation of the printer, Johannes Petreius, for similar advertisements appear on some of his other titles. For example, the introduction to his Bible of 1527 ends:

> *Tu igitur Christiane lector eme, lege, fruere, & in Christo bene vale.*
> Therefore, Christian reader, buy, read, profit & be strong in Christ.

Despite the promise of "the most expeditious tables," Copernicus's magnum opus is not particularly easy to use, and it took Erasmus Reinhold of Wittenberg eight years to formulate the *Prutenicae Tabulae* (which satisfied the title page promise of *De revolutionibus*). Here we will look in detail at how purchasers of Copernicus's book could have struggled to find the position of Mars on the astronomer's birthday, February 19, 1473. (For a modest simplification of this example, we will ignore the assumption that Copernicus was born in the afternoon.)

Finding the celestial position of a planet proceeds in two stages, first to cycle the planet around the heavenly dome, starting from a known position in antiquity, and then to correct that position because the planet rhythmically speeds up and slows down and is observed from a moving earth that also speeds up and slows down.

Around AD 150 the Alexandrian astronomer Claudius Ptolemy faced the same sort of problem, and even though he solved the situation for a fixed earth, his procedure had essentially the same number of steps required in a heliocentric astronomy. Ptolemy's method therefore served as a guide for Copernicus. In both cases there needed to be a root position, called a *radix*, at a specified time or *epoch*, and a *speed*. In *De revolutionibus* these numbers are found in the text, but it might be more accurate to say the radix is *buried* in the text. Here they are for Mars:

Radix: 238° 22'
Epoch: "Incarnation" = midnight 1 January AD 1
Movement per year: 168° 28' 30" 36'''

It would be tedious to tabulate where roughly the planet would be found in its path around the sun, for example, every year for 1,473 years, so here special tables come to the rescue. Ptolemy, requiring a shorter time span, listed the motion for each eighteen-year interval within 810 years, and then a secondary table for eighteen single years. It might seem rather idiosyncratic to select eighteen-year intervals, rather than every ten years, for example, but there is an unexpected reason. In those days books came in two forms, either as scrolls or as codices (like our modern bound books). The *Almagest* was a codex, with a set number of lines per page, and when Ptolemy figured out the total line count per page, forty-eight lines including the headers, everything worked out with the subsidiary tables if he used eighteen-year intervals. (Forty-eight lines worked nicely throughout the entire book, often with two tables per page.) And of course if he needed an interval longer than 810 years, he could simply add together two or more tabular entries.

Meanwhile, around AD 1300 a very clever alternative tabulation was invented in western Europe in what were called the *Alfonsine Tables*, named after the Spanish ruler. (Copernicus himself invested in a copy published in 1492.) This involved what we call the sexagesimal system, which we still use for degrees, minutes, and seconds. Copernicus assumed that his readers would be familiar with this base-60 system, as well as with the concept of the Egyptian year.

In the ancient Egyptian calendar, the year consisted of twelve months, each of thirty days, plus five intercalary days. This totaled 365 days, whereas the Julian calendar, introduced by Julius Caesar, included a leap day every four years. This meant that in 1,460 years the Egyptian calendar, lacking a leap day, slowly cycled through the seasons, and by the time of Copernicus's birth in 1473, the Egyptian calendar was one year and three days ahead of the Julian calendar. As Copernicus states in Book III, chapter 6, "In computing the heavenly motions, however, I shall use Egyptian years everywhere. Among the civil years, they alone are found to be uniform.... For this reason in the computation of uniform motions, Egyptian years are most convenient."

In sexagesimal notation 1,473 is 24, 33; (24 × 60 + 33), but in calculating planetary positions from tables it is customary to use completed years (in this case 1,472) plus the elapsed months and days, or 24, 32; 49 days. To convert to Egyptian years, we add 1 year and 3 days: 24, 33; 52 days.

(Note: To follow the example below in detail, it is desirable to have a copy of *De revolutionibus* at hand. The most readily available, if not the most reliable, translation appears in the *Great Books of the Western World* set, found on the reference shelves of many public libraries.)

We now begin to find the position of Mars on February 19, 1473 (Julian), which in Egyptian years expressed sexagesimally is 24, 34; plus 51 days. In *De revolutionibus* Copernicus gives these movements in a set of tables based on the Egyptian year.

Radix: 238° 22' =	3	58°	22'					
Movement in 24 × 60 Egyptian years	23	24	14	25	= 5	24	14	25
Movement in 33 completed Egyptian years	2	39	40	49				
Movement in 52 days		24	0	7				
Intermediate position of Mars	0	26°	17'	21"				

The power of the sexagesimal system reveals itself in the second line above. Copernicus does not give a second table with sixty-day intervals, because such a table is already embedded in the first. One simply shifts the numbers one sexagesimal place to the left, throwing away the first column because it represents completed movements around the circle. Similarly the leading sexagesimal digit 23 equals $3 \times 6 + 5$, where 3×6 represents three times around the circle, which we can trim away.

We now have a position of Mars with respect to the sun, but what we really want is the position of Mars with respect to the earth, so we need a very similar procedure for locating our own planet. Again, for the sun (which is equivalent to locating the earth), Copernicus gives a pair of tables, which are found in Book III of his treatise. (The Mars tables are in Book V.) The resulting position for the sun is:

Solar Radix: 272° 31' =	4	32°	31'	
Movement in 24 × 60	53	55	38	49" = 5 55 38 49
Egyptian years				
Movement in 33 completed	5	51	39	0
Egyptian years				
Movement in 52 days		51	15	6
Mean longitude of the sun	5	11°	4'	

(These numbers are in the sexagesimal system, so that the leading column is the number of 60° units involved.) Thus 5, 11; 4 equals 311° 4'; this is the calculated position of the sun on Copernicus's birthday, which is 180° from the heliocentric position of the earth.

Ultimately we will need a value for the precession, the slow change of the coordinate system in a cycle now known to be 26,000 years long. *De revolutionibus* presents this in excruciating detail, but it scarcely offers an "owner's manual." Once understood, it is not so difficult because it follows a pattern similar to the two previous calculations. Unlike the preceding calculations, however, to calculate precession requires a secondary table that involves an additional tabular lookup. The starting tables are found in Book III.6. The result from the second or "anomaly lookup" is doubled and entered into a table found in Book III.8, and this result is simply added to the initial result for the basic precession.

Precession Radix:	5°	32'		Anomaly:		6°	45'	
Motion in 24 × 60	20	4	50"		2	30	57	39"
Egyptian years								
Motion in 33		27	36			3	27	34
completed								
Egyptian years								
Motion in 52 days			7'''					54
Basic precession	26°	4'	26"		2	41°	11'	7"
				Doubled:		322°	22'	14"

To finish the computation of precession, the doubled anomaly must be entered in another table, found in Book III.8, which yields 44', and which is added to the precession to get a total of 26° 48'. This number will be used in the final step of the planetary position process.

This completes the first part of the computation, which gives the key to "mean" or average heliocentric positions of the earth and Mars. So far in the calculation, we have worked on the false assumption that Mars moves uniformly in its orbit. The second part, which follows, is to take into account the varying speed of Mars in its eccentrically placed orbit. This part of the calculation requires knowledge of the orientation of Mars's orbit, and for this we need to get the aphelion line. The aphelion is where Mars is farthest from the sun and moving the slowest. Copernicus is shy about giving this position because it continuously changes owing to precession, and he nowhere tabulates it in *De revolutionibus* as a function of time. This omission must have been a serious frustration to early users of his book, but we will work around this absence in the following way. Copernicus analyzed an observation of Mars in June 1523 and gave an updated value of the precession. Because the precession changes the aphelion line rather slowly and the derived final position of Mars is only weakly dependent on the value of the aphelion line, we can use the June 1523 value for his birthday in February 1473.

Not only does Mars move the slowest when at the aphelion, but it will appear to move slower when it is farther from the earth, so both of these effects must be reckoned with.

These calculations could be carried out with mathematical formulas and trigonometrical tables. Or they could be achieved with precalculated tables. For example, we could calculate for every degree of Mars's orbit how much faster (or slower) than average it would be moving. But then we would have to provide such a table for every degree of the earth's position as well. If we made the tables for every degree, that would then be 360 × 360 numbers, or more than 100,000 numbers! Here Ptolemy was a mathematical magician, devising a scheme that worked by linear addition rather than multiplication, and Copernicus followed in his footsteps. In fact, Copernicus went one step further. In Ptolemy's system one had to be very careful as to whether a given correction was to be added to or subtracted from the total. Copernicus fixed it so that it was much more obvious which corrections were additive.

Copernicus called his corrections *prosthaphaereses,* a Greek word taken from Ptolemy meaning "additio-subtractions." His tables are reproduced here from Book V.33 of *De revolutionibus.* The double column at the right of each of the two pages is the argument in degrees (that is, the number one enters into the table), and there are four columns of corrections.

You may wonder if Copernicus also took into account the varying speed of the earth in its eccentrically placed orbit; the answer is no. These deviations for Mars are so much larger than those of the earth that they are simply swallowed up and to a degree amalgamated into the numbers for Mars. It took the work of Johannes Kepler, several generations later, using the massive set of observations from Tycho Brahe, to find and correct this Achilles heel in Copernicus's work.

To proceed with finding the corrections, we must link our derived positions of the sun and Mars with the orientation of Mars's orbit (as given by its aphelion).

λ_{Sun} Sun's longitude	311° 4'
α Mars's position	−26° 17'
λ_a Mars's aphelion	−119° 40'
κ Eccentric anomaly	165° 7'

Here "anomaly" means irregularity or nonuniformity, and "eccentric" refers to an off-centered or eccentrically defined circle. The diagram (Figure 10) shows the position of the eccentric anomaly on

9. The final pages of corrections for calculating a position for Mars, folios 175 verso–176, from the first edition of *De revolutionibus*.

the day of Copernicus's birthday. We use the value of the eccentric anomaly to enter Copernicus's first and second correction columns:

$$c_1(\kappa) = 3° \ 11'$$
$$c_2(\kappa) = 57' \ 55''$$

Now the c_1 correction is added to the Mars position:

$$\alpha = \alpha + c_1 = 26° \ 17' + 3° \ 11' = 29° \ 28'$$

and this number shows the effect of the resulting speed-up as the planet moves from its aphelion toward its perihelion.

The next step makes a much larger correction on account of the changing distance between the earth and Mars as seen from a position along the aphelion line, whereas the final correction compensates for the earth being at a different viewing point:

$$c_3(\alpha) = c_3(29° \ 28') = 11° \ 0'$$
$$c_4(\alpha) = 1° \ 24'$$
$$c_3(\alpha) + c_2(\kappa) \ c_4(\alpha) = 11° \ 0' + (58' \times 1° \ 24'/60) = 12° \ 24'$$
$$\Delta = \alpha - c_3(\alpha) - c_2(\kappa) \ c_4(\alpha) = 29° \ 28' - 12° \ 21' = 17° \ 7'$$
$$\lambda_{Mars} = \lambda_{Sun} - \Delta = 311° \ 4' - 17° \ 7' + (\text{precession}) \ 26° \ 48' = 320° \ 45'.$$

Amazingly, the recipe book (with no organized instructions) has brought up the same answer as the rare Copernican horoscope preserved in the State Library in Munich, 321°. It will not come as a shock to learn that apparently no one except Copernicus's student Rheticus published a set of ephemerides (daily positions of the planets) based directly on Copernicus's work until Erasmus Reinhold's *Prutenic Tables* became available in 1551, eight years after *De revolutionibus* was published. Reinhold recomputed everything to make the tables more complete and easier to use.

Copernicus himself must have appreciated that his were not "the most expeditious tables," and this must have contributed to his reluctance to send his manuscript to a printer. Fortunately, the enthusiastic young astronomer from Wittenberg was persistent and persuasive.

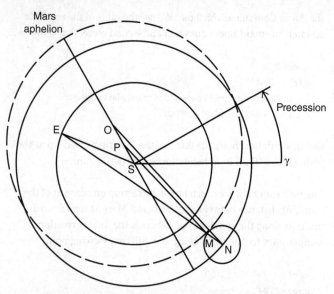

10. **The orbits of Mars and the Earth with the planets positioned for Copernicus's birthday, February 19, 1473. S = the sun, E = Earth, M = Mars. The epicycle is not to scale and has been approximately doubled in size for clarity. Note the long, narrow isosceles trapezoid with equal sides MN and OP. The angle NOP equals κ, the eccentric anomaly, while the angle ESM (not fully completed in the diagram) equals α in the text.**

References

The references listed here to sources for facts and quotations in this short introduction are located by a few words (in bold face) taken from the opening line of the relevant paragraph. These are immediately followed by a few key words identifying the specific position of the citation.

Chapter 1: Copernicus, the young scholar

Nicolaus Copernicus: born in Torun, Poland: the first printed source for Copernicus' birthdate is in Paul Eber, *Calendarium historicum* (Wittenberg: Rhau, 1550); there is in the Bayerische Staatsbibliothek in Munich an early undated manuscript horoscope giving the same date and time.

The printing of books: Sacrobosco's *De sphaera* in 1472: Jürgen Hamel, *Studien zur "Sphaera" des Johannes de Sacrobosco* (*Acta Historia Astronomiae* 51) (Leipzig: Passage-Verl, 2014).

As long as his uncle: returned the document to the notary: Marian Biskup, *Regesta Copernicana* (*Studia Copernicana* VIII) (Wroclaw: Ossolineum, 1973), Register item 33, pp. 41–42.

Meanwhile, near the end: promised to study medicine: Marian Biskup, *Regesta Copernicana* (*Studia Copernicana* VIII) (Wroclaw: Ossolineum, 1973), Register item 38, p. 43.

Chapter 2: The architecture of the heavens

This information about: More than thirty thousand clay tablets: O. Neugebauer, *The Exact Sciences in Antiquity* (Providence:

Brown University Press, 1957), chap. V, "Babylonian Astronomy," 97–144; O. Neugebauer, *A History of Ancient Mathematical Astronomy* (New York: Springer, 1975), 347–53.

Ptolemy's *Mathematike syntaxis*: Ptolemy's *Mathematike Syntaxis*: G. J. Toomer, translator and annotator, *Ptolemy's Almagest* (Princeton: Princeton University Press, 1998).

Copernicus, in a different age: The divine handiwork: Copernicus, *De revolutionibus orbium caelestium* (Nuremberg: Johann Petreius, 1543), f. 10.

Chapter 3: Copernicus's vision

The same effect occurs: by a golden chain: Rheticus, *Narratio prima* (Danzig, 1540); for an English translation, see Edward Rosen, *Three Copernican Treatises* (New York: Octagon, 1971), 164–65.

Copernicus's vision: harder to walk west than to walk east: Johannes Schöner, *Opusculum geographicum* (Nuremberg: Johann Petreius, 1533), f. A3v, quoting Regiomontanus, "Patet qua si sic, difficilius esset ire contra occidentem quae orientem, quod est contra experientem."

The emphasis is on: obscure manuscripts were found: Bernard Goldstein, "The Arabic Version of Ptolemy's *Planetary Hypotheses*," *Transactions of the American Philosophical Society* 57, part 4 (Philadelphia: American Philosophical Society, 1967).

Unfortunately, Copernicus: reference to Aristarchus: Owen Gingerich, "Did Copernicus Owe a Debt to Aristarchus?" *Journal for the History of Astronomy* 16 (1985): 36–42.

But none of the vague: "In no other way do we find: Copernicus, *De revolutionibus orbium caelestium* (Nuremberg: Johann Petreius, 1543), f. 10.

Chapter 4: Canon days and the *Little Commentary*

It was now time: *The Little Commentary*: see Noel M. Swedlow, "The Derivation and First Draft of Copernicus's Planetary Theory: A Translation of the Commentariolus with Commentary," *Proceedings of the American Philosophical Society* 117.6 (1973): 423–512. The part of the text quoted here is on pp. 433–36 and 510 and has been somewhat edited. Also consulted was the translation by Edward Rosen, pp. 81–82 in *Nicholas Copernicus Complete Works III*, Pawel Czartoryski, ed. (Warsaw-Cracow: Polish Scientific Publishers, 1985).

Calippus and Eudoxus: concentric circles in no way permit: besides the basic eccentric deferent, two small epicyclets were used to replace the equant, plus two more circles to account for the planetary latitudes for Venus, Mars, Jupiter, and Saturn. Because neither Ptolemy nor Copernicus could observe Mercury at certain critical times, the model was overly complicated and was not corrected until Kepler, using Tycho's numerous data, made Mercury's model congruent with the other planets.

Many twentieth-century scholars: "cycle on epicycle, orb on orb": John F. W. Herschel, *A Preliminary Discourse on the Study of Natural Philosophy* (London: Longman, 1831), 266.

Precisely when did Copernicus: It is hard to believe: Swerdlow, ibid., p. 431. This concerns the so-called homocentric spheres, mentioned in Chapter 2, which does not allow a changing distance from the earth to account for a variable brightness of a planet.

Although the details: hissed off the stage: Owen Gingerich, *The Book Nobody Read* (New York: Walker & Co., 2004), 135.

Chapter 5: Competing with Ptolemy

Two additional complex: he described them in his own book: see *Tycho Brahe's Description of his Instruments and Scientific Work as given in Astronomiae Instauratae Mechanica (Wandesburgi 1598)* (Copenhagen: København: I Kommission hos E. Munksgaard, 1946), 44–47. Tycho Brahe published the poem earlier in his *Epistolarum astronomicarum* (Uraniburg, 1596), 235–36.

Regiomontanus, who was: something appreciably wrong: *Scripta clarissimi mathematici M. Joannis Regiomontani* (Norimbergae, 1544), f. 42.

Chapter 6: Rheticus

Meanwhile Luther's Reformation: Alexander Scultetus: Edward Rosen, "Nicholas Copernicus: A Biography," in *Three Copernican Treatises* (New York: Octagon Books, 1971), 383–85; 395–97.

The prologue begins: and beheaded: "Given his crime, he should have been hanged. But because he was a Bürger of Feldkirch and no mere common criminal, he was granted instead a privilege reserved for condemned citizens since the times of the Romans: He was publicly beheaded with a sword." Dennis Danielson, *The First Copernican* (New York: Walker & Co., 2006), 17.

Philipp Melanchthon: rebellious satirical poet: for Lemnius the "skilled poet with a wicked eye for satire," see pp. 26–30 in Dennis Danielson, *The First Copernican* (New York: Walker & Co., 2006).

Together with a young: progress report to Schöner: the letter to Schöner, mentioned on the first page of Rheticus' *Narratio prima*, has not survived; for an English translation, see Edward Rosen, *Three Copernican Treatises* (New York: Octagon Books, 1971), 111.

Eventually Rheticus: "rarely allowed us to see Mercury: see "Ptolemy and the Maverick Motion of Mercury," selection 5 in Owen Gingerich, *The Great Copernicus Chase and Other Adventures in Astronomical History* (Cambridge: Sky, 1992), 31–35.

Still Copernicus hesitated: He confided to Rheticus: Georg Joachim Rheticus, *Ephemerides novae* (Leipzig: Passage-Verl, 1550), f. A3v.

"Nature does nothing: this one motion of the earth: Rheticus, *Narratio prima* (Danzig, 1540); for an English translation, see Edward Rosen, *Three Copernican Treatises* (New York: Octagon, 1971), 137.

It was a polemic: *by a golden chain*: ibid., 164–65.

Chapter 7: *De revolutionibus*

[last paragraph]: pages were sent to Copernicus: see the discussion under "Rheticus and the Extended Errata List," in Owen Gingerich, *An Annotated Census of Copernicus' De Revolutionibus* (Leiden: Brill, 2002), xviii–xix.

Chapter 8: The book nobody read

There was something else: mental powers had abandoned him: Dennis Danielson, *The First Copernican* (New York: Walker & Co., 2006), 2.

But both Giese: This introduction read: the text of the Andreas Osiander *Ad lectorem* is here translated and edited by Owen Gingerich.

Gradually a few people: Maestlin wrote: Owen Gingerich, *An Annotated Census of Copernicus' De Revolutionibus* (Leiden: Brill, 2002), 220–21.

How did Maestlin: showed his book to his teacher: Owen Gingerich, *The Book Nobody Read* (New York: Walker & Co., 2004), 163–65.

Among the twentieth-century: Arthur Koestler: Arthur Koestler, *The Sleepwalkers* (London: Hutchinson, 1959), 191.

In 1970, as we: I found a first edition: Owen Gingerich, *The Book Nobody Read* (New York: Walker & Co., 2004), 22–25.

There is a well-known: Wittich owned and annotated: Owen Gingerich and Robert S. Westman, "The Wittich Connection: Conflict and

Priority in Late Sixteenth-Century Cosmology," *Transactions of the American Philosophical Society* 78, part 7 (1988), 27–42.

An early mention of: *The Castle of Knowledge*: Robert Recorde, *The Castle of Knowledge* (London, 1556), 165.

Earlier celestial images: Digges's own copy: Owen Gingerich, *An Annotated Census of Copernicus' De Revolutionibus* (Leiden: Brill, 2002), 215.

Meanwhile, in 1609: converted into a crusading heliocentrist: Owen Gingerich and Albert Van Helden, "How Galileo Constructed the Moons of Jupiter," *Journal for the History of Astronomy* 42 (2011): 259–64.

As a consequence: copies in Italy were censored: Owen Gingerich, *An Annotated Census of Copernicus' De Revolutionibus* (Leiden: Brill, 2002), xxiii–xxv, 367–68. *Dialogue on the Two Great World Systems: Dialogo sopra i due massimi sistemi del mondo tolemaico, e copernicano*, Galileo Galilei, Florence, 1632.

As part of my: search for copies: these five paragraphs are closely paraphrased from the epilogue of Owen Gingerich, *The Book Nobody Read* (New York: Walker & Co., 2004).

In two summer sessions: Gąssowski's team found thirteen burials: Jerzy Gąssowski and Beata Jurkiewicz, "The Search for Nicolaus Copernicus's Tomb," in Jerzy Gąssowski, ed., *The Search for Nicolaus Copernicus's Tomb* (Pułtusk: Pułtusk Academy of Humanities, 2006), 9–19.

Then, when in a lecture: mitochondrial DNA: Owen Gingerich, "In the Orbit of Copernicus," *American Scholar* 80.3 (Summer 2011): 43–49; Owen Gingerich, "The Copernicus Grave Mystery," *Proceedings of the National Academy of Science* 106 (2009): 12215–16.

Further reading

Copernicus, Nicolaus. *On the Revolutions of the Heavenly Spheres.*
Translated by Charles Glenn Wallis. Great Minds Series. Amherst,
NY: Prometheus Books, 1995. This translation appeared in *Great
Books of the Western World*, Chicago, 1952, vol. 16, 2nd ed. 1990,
vol. 15.

Danielson, Denis. *The First Copernican: Georg Joachim Rheticus and
the Rise of the Copernican Revolution.* New York: Walker & Co.,
2006.

Ferguson, Kitty. *Tycho & Kepler: The Unlikely Partnership That
Forever Changed Our Understanding of the Heavens.* New York:
Walker & Co., 2002.

Galilei, Galileo. *Sidereus Nuncius, or The Sidereal Messenger.*
Translated with introduction by Albert Van Helden. Chicago:
University of Chicago Press, 2016.

Gingerich, Owen. *The Book Nobody Read: Chasing the Revolutions of
Nicolaus Copernicus.* New York: Walker & Co., 2004.

Gingerich, Owen. *The Eye of Heaven: Ptolemy, Copernicus, Kepler.*
New York: American Institute of Physics, 1993.

Gingerich, Owen. "From Copernicus to Kepler: Heliocentrism as
Model and Reality." *Proceedings of the American Philosophical
Society* 117.6 (1973): 513–22.

Koyré, Alexandre. *The Astronomical Revolution.* Ithaca, NY: Cornell
University Press, 1973.

Repcheck, Jack. *Copernicus' Secret.* New York: Simon & Schuster,
2007.

Rosen, Edward. *Three Copernican Treatises.* New York: Octagon
Books, 1971.

Sobel, Dava. *A More Perfect Heaven*. New York: Walker & Co., 2011.

Swerdlow, Noel. "The Derivation and First Draft of Copernicus's Planetary Theory." Translation of the *Commentariolus* with Commentary. *Proceedings of the American Philosophical Society* 117.6 (1973): 423–512.

Swerdlow, N. M., and O. Neugebauer, *Mathematical Astronomy in Copernicus's De Revolutionibus*. New York: Springer-Verlag, 1984.

Voelkel, James R. *Johannes Kepler and the New Astronomy*. New York: Oxford University Press, 1999.

Index

Copernicus